FOUR GERMANYS

The Shortest
History of
the

Hawes

[signature]

In the series *Politics, History, and Social Change,*
edited by John C. Torpey

FOUR GERMANYS

A Chronicle of the Schorcht Family

DONALD S. PITKIN

Foreword by JOHN C. TORPEY

TEMPLE UNIVERSITY PRESS
Philadelphia • Rome • Tokyo

TEMPLE UNIVERSITY PRESS
Philadelphia, Pennsylvania 19122
www.temple.edu/tempress

Copyright © 2016 by Temple University—Of The Commonwealth System
   of Higher Education
All rights reserved
Published 2016

Library of Congress Cataloging-in-Publication Data

Names: Pitkin, Donald S. (Donald Stevenson), 1922–2012, author.
Title: Four Germanys : a chronicle of the Schorcht family / Donald S. Pitkin ;
   foreword by John C. Torpey.
Description: Philadelphia, Pennsylvania : Temple University Press, 2016. |
   Series: Politics, history, and social change | Includes bibliographical
   references and index.
Identifiers: LCCN 2016005350 (print) | LCCN 2016026183 (ebook)
   | ISBN 9781439913420 (cloth : alk. paper) | ISBN 9781439913437
   (paper : alk. paper) | ISBN 9781439913444 (ebook)
Subjects: LCSH: Schorcht, Edwin, 1890–1976—Family. | Thuringia
   (Germany)—Social life and customs—20th century. | Thuringia
   (Germany)—Social conditions—20th century. | World War, 1939–1945—Social
   aspects—Germany—Thuringia. | Farmers—Germany—Thuringia—Biography.
   | Thuringia (Germany)—Biography.
Classification: LCC DD801.T48 P58 2016 (print) | LCC DD801.T48 (ebook) | DDC
   929.20943—dc23
LC record available at https://lccn.loc.gov/2016005350

♾ The paper used in this publication meets the requirements of the American National
Standard for Information Sciences—Permanence of Paper for Printed Library Materials,
ANSI Z39.48-1992

Printed in the United States of America

9 8 7 6 5 4 3 2 1

For my sister, Jane

Contents

Foreword

ONALD S. PITKIN researched and wrote *Four Germanys* after he had retired from a long career of teaching—first at Northeastern University and then, subsequently, at Amherst College. It is a book much like the one that he had written previously, *The House that Giacomo Built*, which details the triumphs and travails of an Italian peasant family over an extended historical period. In addition to the book, that project also resulted in a film by the same name. The reader of this book will immediately see the cinematic potential here as well. Indeed, one is struck by the way the two books form part of a larger whole, a diptych of twentieth-century European life.

Although Don was an anthropologist by training, the book exemplifies C. Wright Mills's oft-quoted dictum that sociology is the marriage of biography and history. That is, a work of social "science" should bring together both the historical conditions in which people act and their experiences of and responses to those conditions. *Four Germanys* does this superbly: one gets the Schorcht family's "ground-level" view of life in imperial Germany, the Weimar Republic, the Nazi period, communist-ruled East Germany, and, finally, the post-unification period. On the basis of archival records and extensive interviews with a number of the living members of the family, Don was able to reconstruct their lives in extraordinary detail: the wedding dress of a young bride in the early 1920s, the varied reactions of the parties to the transfer of the Schorcht family property, the youthful pride of a "Young Pioneer" in the German Democratic Republic period, the ambivalence of the young woman raised in communist East Germany to the promise and uncertainty of life in unified Germany. The constraints imposed by political conditions are sketched in deftly rather than portentously, and moral(istic) judgments about the behavior of family members are handled with a light touch. The author has his views about the broader political developments but understands that those developments may have confronted the members of a rural farm family differently from the way they confronted an outside observer. On the other hand, as a participant in World War II, Don was far from a disinterested outsider: he was aware that he was (as, indeed, his father was) on the other side of a conflict in which members of the Schorcht family were combatants and that a different outcome of the military confrontation could have had momentous consequences for him and the entire world.

Whereas it has now become common for anthropologists to study the "developed" world, when Don first began his career, this was novel and hence, in career terms, risky. But this book is a reminder of the unity of the social sciences and history, a perspective that is losing traction with the advance of academic specialization. It is also a model of writerly craftsmanship; it is virtually without academic jargon and hence completely accessible to the nonspecialist. Don begins the book with a few reflections on the relationship between history and ethnography—on how they fit together and diverge—but then that is that. The rest is a detailed portrait of the Schorcht family and its vicissitudes across the familiar series of twentieth-century German political regimes. Reading the book permits one to understand how Don Pitkin might have enjoyed and approached the job he faced upon arriving at Amherst College: creating a joint Department of Anthropology and Sociology. The writing probably also reflects the demands of teaching undergraduates (who, though curious, have yet to become swamped in academic jargon) rather than graduate students.

It should be noted that the manuscript was essentially completed some two decades ago, long before Don started to become seriously ill. As a result, it does not reflect all the subsequent scholarship on twentieth-century Germany and may therefore appear to scholars to be somewhat out of date. But this is a flaw of little consequence. *Four Germanys* makes an outstanding contribution to our understanding of recent German life. It stands in the traditions of oral history and *Alltagsgeschichte* ("the history of everyday life"), which brought anthropological methods into the study of history. The book displays a remarkable knowledge of life on a farm, the use of farm equipment, and the rhythms of farm life. And it sharpens our understanding of the relationship between biography and history. The publication of this volume is, I hope, a fitting tribute to the scholarly career of Donald S. Pitkin and a small gesture of appreciation on my part for the pleasure of having been his student and friend.

John C. Torpey
Graduate Center, City University of New York

Acknowledgments

WE ARE HONORED that Temple University Press is posthumously publishing our father's manuscript. This is the culmination of his research and dedication in the years following his retirement from teaching at Amherst College.

We thank the many people who helped Don in the research, discussions, explorations, collaboration, and assemblage of all the information, including Stephen Kampmeier, Andreas Westerwinter, Anke Bahl, Tina Dirndörfer, and Torsten Kolbeck, his research partners in Weimar and Göttern, Germany, who worked diligently on the primary evidence. We also acknowledge John Eidson, for making so many key introductions, and Amy Johnson, Don's dear friend and travel companion.

On the home front, thanks go to Jay Venezia, Mike Kerber, Donovan Cox, and Sean Bethard for their daily support in helping organize the manuscript and to Jamie Nan Thaman for her initial editing work. We are also grateful to Don's niece, Katherine Donahue, a fellow anthropologist, for being a torchbearer and for establishing the relationship with John Torpey and Temple University Press.

Many thanks are due to Joan Vidal and Heather Wilcox for conducting the final review of the book.

Most importantly, our heartfelt thanks go to the Schorcht family for their candor, patience, understanding, and willingness to participate in this endeavor.

Roxie Pitkin Permuth and Steve Pitkin

FOUR GERMANYS

Prologue

Locating Weimar, Göttern, and a Family

ON MY WAY OUT of Berlin on August 4, 1991, en route to Leipzig, I decided that Weimar was where I would make a study of an East German family. Ever since November 1989, when the wall came down, I had been thinking of doing anthropological work in the former German Democratic Republic (GDR). I was interested by the singular opportunity it offered to investigate the lives of persons who had been witness to, and actors in, various moments in the sequence of political, social, and economic changes that had occurred over the last seventy-five years of East German history: the Weimar Republic, the Third Reich, the GDR, and now a united Germany. It was my opinion that looking at the history of a family, an "ordinary" family, offered a fruitful way of locating people in a setting to reflect the impact of critical historical events while also revealing how one particular family managed its own historical adaptation to these events.

I walked and drove through Weimar for two days. In the end, it was the actual physical relationship of the German National Theater, where the constitution of 1919 had been approved, the houses and museums of Johann Wolfgang von Goethe and Friedrich Schiller, and the Buchenwald camp—set in the beech forest on the northern outskirts of the city, where Goethe and Charlotte von Stein had picnicked—that brought home to me the degree to which Weimar encapsulated within its own history the contradictions of modern German history. It seemed to me that Weimar and its surroundings—at the confluence of much that, for better or worse, was essentially German—would be an appropriate site from which to carry out my study. The identification of a family, however, would have to wait another several months.

The school year 1991–1992 was the last I taught at Amherst College, where I had been a professor of anthropology since 1964. Plans for carrying forth what I had, by then, named the Weimar Family Project were laid aside until the mid-year break, when I returned to Germany. After a few days in Berlin, I traveled to Leipzig to make contact with a 1990 Amherst graduate, Stephen Kampmeier, then a student of German literature at the University of Leipzig. I had written to Stephen of my desire to identify a family to research in Weimar, and he had agreed to help me in any way possible.

After meeting up with Stephen, we began our drive to Weimar, stopping to spend the night in Jena, some 13 miles to the southeast of Weimar, to meet with Terry May, a professor of German at St. John Fisher College in Rochester, New York. May was in Jena for the year teaching American studies. As he was knowledgeable about Weimar, I used the occasion to speak to him about my expectations for the Weimar Family Project and how best to go about identifying a family for study.

As chance would have it, May knew of a family whom he thought might be appropriate, as he knew them to be particularly forthcoming and hospitable. They lived in the small town of Göttern, some 15 kilometers to the southeast of Weimar but still within the county limits. In fact, he happened to be going to a birthday party at their house that afternoon and would gladly introduce us if we would be so kind as to give him a ride. When I readily accepted, he telephoned ahead to say that we were coming. I was delighted and surprised at the prospect of locating a family so soon and yet disconcerted by the fact that the family was so far from Weimar. I was not yet ready to scuttle my plan of researching a family in the city.

On the way to Göttern, Professor May told us about the Schorcht family and how he had come to know them through Heidrun Schorcht, an assistant professor of American studies at the University of Jena. The Schorchts, once a farming family, had been unable to resume private farming after *die Wende* (literally "turn," referring to the fall of the GDR regime) due to the dislocations created by the enforced collectivization program of the 1960s and 1970s; at that time, however, farmers' children were afforded many new opportunities as a result of industrialization and higher education.

When we arrived in Göttern, I felt favorably disposed to the family we were about to meet—all the more so by my captivation over the great wooden gate that opened to allow us to proceed into the courtyard, around which the farmstead was built. I felt as though I had been admitted into a different world, distant in time and place from the once-industrialized countryside around us.

A sense of the possible readiness of the Schorchts to entertain the idea of being subjects of a book was evident in their warm welcome of two strangers to partake in a family celebration. We quickly became conversant with the family members present: Edgar Schorcht (born 1922), head of the family and retired farmer; Wally, Edgar's wife (born 1926); Erhard, their older son (born 1951); Heidrun, Erhard's wife (born 1952); Maria, Erhard and Heidrun's thirteen-year-old daughter; Roswitha Netz (born 1959), Edgar and Wally's daughter; Manfred, Roswitha's husband (born 1956), whose birthday it was; and Roswitha and Manfred's two children, Katherina (thirteen) and Philipp (six). I gathered that both Erhard and Roswitha and their respective families lived with their parents in the converted homestead, while the other son, Norbert (born 1955); his wife, Renate (born 1957); and their three children, Magdalena (fifteen), Julia (eleven), and Anna (two), lived in a nearby town.

We were plied with *Kaffee und Kuchen* in honor of the celebrant's birthday. Later, when the dishes were cleared and Professor May had cued the family as to the principal reason for my visit, we began to talk. I told them of my hope of finding a three-generation family willing to work with me in reconstructing its history in the form of a book, to inform others of how its members had negotiated the radical swings in the political and economic fortunes of recent eastern German history. I added that I hoped they might consider being that family.

That I had touched on a responsive chord seemed evident when Edgar Schorcht brought forth a family photo album, as if to say that it was none too soon to begin our work together. The album was largely devoted to pictures of his father, Edwin, a veteran of Verdun. Edwin had assumed control of the family farm from his father, Hugo, in 1922. He, in turn, had signed the farm over to Edgar in 1967. Officially retired, Edwin had continued to live on the farm with Edgar and his family until his death in 1976. His wife, Elly—Edgar's mother—had died several years before Edwin's passing.

It was becoming clear that if I were to have the good fortune of working with the Schorcht family, I would have Edgar and Wally Schorcht's resources in helping bring Edwin and Elly to life at the moment the farm became theirs, when the Weimar Republic was still in its infancy. I learned that Edgar, like me, was born in 1922. He came of age in the Nazi period and was a member of the Hitler Youth. He had fought on the Russian Front while I was engaged in the war in the Pacific. In April 1945, Edgar was taken prisoner by the Americans. When the Russians took control of Thuringia from the Americans in July 1945, the Schorchts entered into the long encounter with Communist hegemony that resulted in the partial and then later complete collectivization of the farm by 1971.

I learned that Erhard, Heidrun, Roswitha, Manfred, Norbert, and Renate, all born in the 1950s, were at first Young Pioneers (members of the state-run youth organization) and later members of the Free German Youth, or *Freie Deutsche Jugend* (FDJ), as the regime attempted to make of them "well-rounded socialist personalities." Like everyone else, they were taken unaware by the sudden turn of events of November 1989, ushering in yet another Germany, which their children know to be their own.

By the time it became opportune to bid each other good-bye, I left tired yet privileged to have made contact with a family so hospitable and so willing to share their experiences and memories; it boded well for a possible collaboration in the future. I told them I would write when I returned to the United States, allowing them time for family deliberation.

By the time we arrived in Weimar later that evening, it was clear to me that I would be well served in letting go of the idea of studying an urban Weimar family whose history would be interwoven with that of the city. Instead, I would study a family living in the countryside, where socialist policies were radical in their intention. Such a strategic shift in the focus of the project would depend, of course, on the assent of the Schorchts. I was confident, however, that they would agree. Nevertheless, the wealth of archival and other resources pertaining to Weimar as the administrative capital of the district in which Göttern lies would undoubtedly continue to prove invaluable to me, just as the city's own rich history would inspire in me a respect for historical detail. The project would, I envisioned, be headquartered in Weimar, with trips to Göttern for interviews. As one condition of working together, I had assured the family that I would be as minimally intrusive on their privacy as possible.

In the next two days, we made contact with the mayor of Weimar, the director of cultural affairs, and those in charge of the relevant libraries and archives, while obtaining information about the possibility of living arrangements.

When I left Germany toward the end of January 1992, I was optimistic that the Weimar Family Project would, in fact, get under way as planned. What remained was to complete my last semester of teaching at Amherst College, apply for project funding, and prepare to return to Germany by July so that the research and my retirement could go hand-in-hand.

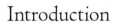

Introduction

Family

THIS BOOK follows in the footsteps of its predecessor, *The House that Giacomo Built: History of an Italian Family, 1898–1978*. I have written both of these books with the understanding that what we call "family" provides an institutional context for referring to processes that are both personal and social. It is this strategic location of family as a point of conjunction between what is specific to the individual and what is general to the community that makes it a fruitful subject of investigation. Family research offers a way into mentalities and a window on the larger world.

Yet understanding of familial existence inevitably begins with lived experiences of specific families, whether our own or others. Only later do we make the cognitive leap to the inferred category of family: the Italian family, the German family, the Western family, and so on. To assign a greater truth value to the existence of any one family, however, as opposed to family in general is not to make the essentialist assumption that actual families are a given in nature—they are, like their more abstract counterpart, social constructs, even though they are constructed at a lower level of inference.

Narration

The writing of a family history involves the telling of a story—that is, narration, or transformation, to paraphrase Hayden White, of a sequence of events into a discourse, a story, an "emplotment."[1] Telling a story about the way things seem to have been requires transforming those events into an intelligible representation, for the real world is not represented in the form of well-made stories—it does not come to us already narrated. The intelligibility that the representation of reality requires necessitates the imposition of imagination on the part of the narrator. The alternative would be a mere list of dates and events that would not constitute the "realism" that we demand of history. As the author of a history of the Schorchts of Göttern, I am more than a passive chronicler of their history. I am the creator of that history, establishing its contours and interpretive direction. This involves selecting, from among the wealth of information I am privy to, what merits inclusion as family history and what deserves omission. I have made judgments in regard to inclusion/exclusion at every step along the way. In the best of all possible worlds, the reader would be made privy to that ongoing process, but the resulting cumbersomeness would be intolerable.

What I am saying in respect to the authorship of historical texts is nothing more or less than what Clifford Geertz has stated in regard to ethnographic authorship: "that all ethnographical descriptions are homemade . . . are the describer's descriptions, not those of the described."[2]

History, Memory

The narration of a family history entails moving beyond its confines to engage in the multiple stories that encompass it, if only because family members assume roles outside as well as in the household. The identification of this intersection of histories—personal, familial, regional, and national—lies at the heart of what C. Wright Mills has called "the sociological imagination." He writes, "No social study that does not come back to the problems of biography, of history, and of their intersections within a society has completed its intellectual journey."[3] In a study that is in large part oral history, biography appears as subjective memory. Memory, however, is of problematic value for recovering the past, even though it constitutes important testimony about it. Its importance lies not in being more authentic or "truer" than history but in suggesting how the past was experienced. Thus, memories are not enough; history must be there as well, "to transform them," as Jacques Le Goff writes, "into something that can be conceived, to make them knowable."[4]

In the case of German history, and particularly that of the Third Reich, the relationship between history and memory is special. As the Jewish historiographer Yosef Yerushalmi writes, "Memory and modern historiography stand, by their nature, in radically different relations to the past."[5] On the one hand, "Jewish" memory is "drastically selective" in its unrelenting focus on the Holocaust as "the central event in modern German Culture."[6] On the other hand, "German" memory is, as Saul Friedländer points out, characterized by various degrees of avoidance and denial in its representation of the Nazi epoch.[7]

As the narrator of the history of a German family that includes twelve years of Nazism in its scope, I assumed that questions and issues, especially those pertaining to National Socialism, might well be treated like skeletons in the closet, better left in the dark. The obvious queries had to do with the Nazi Party and membership in it, Adolf Hitler, the war, the Jews, the Holocaust, and knowledge about the nearby Buchenwald camp—all profoundly moral questions. As I began working, I intuited a hidden agenda at work when these topics came up—an inevitable residue of the war in which Edgar and I had been mortal enemies. I sensed an expectation on the part of the Schorchts that I, as the "other," would want to know, above all else, about those matters and their involvement in them. The anticipation of judgments leveled against a people proud of their heritage kept these topics away from the center of our concerns. That they were more peripheral than central could have been due to the degree to which they were engaged with them at the time. Yet I knew that this was an issue that I could not readily resolve.

I remember driving back to Weimar after the first interview with the Schorchts, noting the place in the road between Magdala and Mellingen where the massive bell tower monument at Buchenwald becomes visible. The sight of the tower, built in 1955, had no relevance to the question of whether the Schorchts were aware of what went on there—in actuality, the camp was hidden from view by the crest of the Ettersberg hill—but it exists as evidence of a history that will not go away.[8]

Alltagsgeschichte

Much of family life is not complicit with large historical events but revolves around the responsibilities of everyday life, *Alltag*. The history of everyday life, *Alltagsgeschichte*, became an important new undertaking in West German historiography in the mid-1970s, encompassing as it did a wide range of historical initiatives regarding the lives of ordinary people as enacted in family life, in gender relations, in the workplace, and so on, to represent "history from below." The emergence of *Alltagsgeschichte* signaled the inclusion of an anthropological approach in German social history in its emphasis on the experiential. As Geoff Eley writes, "This turn toward ethnology also involved a shift in the historian's agenda from impersonal social process to the experiences of human actors."[9]

As an American anthropologist, I welcome the convergence of anthropology with history that *Alltagsgeschichte* represents and the opportunity it offers to work under its banner. At the same time, I recognize that undertaking a study lying within its parameters bears its own set of responsibilities. The most serious concern is the danger that historical attention paid to the affairs of the everyday may normalize that deemed abnormal—in this case, the Third Reich. *Alltagsgeschichte* may elicit relativism by revealing that behavior and ideas bearing the stigma of evil in one particular historical circumstance inevitably have properties in common with the "normal" in other contexts. This is not to suggest that the danger of relativism is to be avoided by not subjecting the Nazi period to the focus of the history of everyday life but to recognize that the sources of evil are to be sought in the unexamined intentions of what is assumed to be normal. To think otherwise is to essentialize Nazism and more particularly the Holocaust at the cost of deferring their ultimate understanding.[10] More concretely, to attribute historical status to the events of everyday life is to suggest the existence of a layered imagery of historical process in which political and ideological activity overlies a quotidian culture made accessible by historical ethnography. What is instructive about such a duality is that "below the barbarism and the horror of the regime," as Ian Kershaw points out, "were patterns of social 'normality' that were, of course, affected by Nazism in various ways but that predated and survived it."[11] In fact, the life course of both Edgar and Wally Schorcht and their "everyday" existence, molded for the most part by non-ideological considerations, cuts a swath through the political and ideological fluctuations of four Germanys. The essential uniformity of their lives consisted of strands of behavior and thought whose basic properties remained effectually constant but whose surfaces were prone to elaboration and transformation. Indeed, the relativistic perspective that a longitudinal study of the Schorcht family provides suggests that the same fealty to the doctrines of the Third Reich and later to the GDR was to be found in a prior allegiance to secular authority.

Yet it is far from my purpose to draw attention to the Third Reich—so brief in duration, so murderous in intention—over that of the divided Germany that followed it.

Communalism

In the period subsequent to the defeat of the Nazi regime in 1945, the people of eastern Germany became citizens of yet another Germany, the socialist German Democratic Republic (GDR). From its beginning, the GDR was an unintended political entity. Its inception was unforeseen by the victors as well as by German politicians during and immediately after the cessation of hostilities. The precedent for a division of Germany had been established in the demarcation zones of occupation established first at Yalta and later at Potsdam.[12] In default of agreement on the governance of Germany among the allies, the Western powers established the Federal Republic of Germany.

In response, the Soviet Union proclaimed the German Democratic Republic. Founded on October 7, 1949, the GDR assumed the eradication of Fascism and the creation of an advanced socialist society as its historic task.

Even before the GDR came into being in 1949, the Soviet Military Administration in 1945 presumed upon the German leaders in the Soviet zone of occupation to launch a land reform program, designed to remove the "cancer" of Junkerism, perceived as a promulgator of Fascist adventurism. In its wake, land reform brought about a profound change in the power structure of East Germany.

Soon to follow in 1952 was a program of social transformation even more radical in its goals: the audacious collectivization of agriculture and the industrialization of the countryside, intended to bring into being a new person, the socialist agrarian worker. The extent to which the East German regime, in carrying forth its vision, was indeed successful in instilling within the people a "socialist" conscience—substituting "we" for "I"—was due in part to the survival of the National Socialist ethos of *Volksgemeinschaft* (national community), the idealization of the common good.[13] Stephen Fritz effectively makes the point that the communitarian aspect of National Socialism, inherent in *Volksgemeinschaft*, had explosive appeal, especially among younger people. He writes, "The allure of Nazism, then, lay in its creation of the belief that one was in service to an ideal community which promoted both social commitment and integration."[14]

It is perhaps ironic that those two Germanys, that of Hitler and that of Walter Ulbricht, left behind a legacy of communalism on the borders of Western Europe. It is a legacy that members of the Schorcht family have given their allegiance to at different times, in different ways, and to different degrees.

Individualism

With the pronouncement of unification on October 3, 1990, the people of the former GDR became citizens of the amplified Federal Republic of Germany. Immediately, winds of change blowing from the West brought about the valorization of the individual and consumption in place of adherence to the community and to allocation.

In fact, however, the independent enterprise has not returned in every sector of the economy in the "new" German states (*Länder*), as it was assumed by many that it would. The family farm has all but disappeared. High start-up costs, the need for farms of large hectarage, the legacy of the extended fields, and the collectivized infrastructure have selected against the reemergence in the East of the independent farmer. In the case of the Schorchts, their land is now sown, cultivated, and harvested by what had once been a party-controlled collective, now operating as a privately controlled cooperative. After six generations of residence in Göttern, the Schorchts have ceased to be farmers.

Edgar and Wally Schorcht, now retired, enjoy their newly acquired freedom to travel. They can readily visit Edgar's sister, Irmgard, in the West without a visa or join others their age on a bus trip to the Italian lakes. They appreciate the comfort and convenience that goods not previously available bring to them. Nevertheless, they recall, almost daily, the absence of something they have missed since unification: being in the company of others in the pursuit of a common purpose.

Erhard, Heidrun, and Maria are all children of the GDR. Maria was twelve when unification occurred. Erhard, a freelance photographer, has worked in the free market since shortly after the Wall came down. He is, among other things, a regular contributor to the *Neue Thüringer Illustrierte*. Erhard is impressed with how effective the capital economy is in generating productivity when compared with the planned economy he grew up with. On the other hand, he is concerned with its social costs—the anxiety he has noted in those around him faced with the threat of layoffs. He is gratified that Heidrun's job pays well.

Heidrun was relieved and saddened by the collapse of the GDR—relieved that the cynicism and the obdurate authoritarianism of its government was over with, saddened that her strongly felt hope for a viable socialist society had been extinguished.

Born and raised on a small farm, Heidrun was gifted with abilities that were encouraged by the authorities. After being introduced to American literature in high school, she was accepted to the University of Jena, where she earned a diploma in language studies and a doctorate in American studies. Now Heidrun teaches German in the Lithuanian provincial capital of Panevėžys. She is proud of her ability as an independent woman to adapt herself to Lithuanian culture, just as she did in the United States in 1979 in the course of gathering material for her dissertation. Heidrun is confident that she can meet the new challenges flowing from the West, a confidence nurtured by her upbringing in the East.

Method

In conclusion, allow me to point out how I, in the company of my assistants, proceeded to gather data. Unlike in Italy, where an informal atmosphere reigned, a certain formality was the order of the day in Germany as concerned fieldwork. Appointments for a morning or an afternoon interview were made by telephone. Arriving promptly in Göttern from Weimar in my 1971 Russian Lada, we were met at the door by Frau Schorcht. I presented her with flowers, from which I first removed the paper wrapping, as custom requires.

We were then joined in the small sitting room by Herr Schorcht. After a half hour devoted to sharing Wally's coffee and cake, the tape recorder was started, beginning the ninety-minute interview.

Edgar and Wally chose to be interviewed only together. Edgar was a magnificent respondent, a generous narrator, while Wally listened with seeming interest to much of what she must have heard on numerous prior occasions. From time to time, she interjected her own comments or called her husband to task on a question of fact. Edgar may have been selective about what he was ready to comment on. Questions about agriculture and the homestead were answered in detail; those concerning the Nazi period were responded to less generously than were queries regarding the communist era that followed. Edgar was only eleven when the Nazis came to power but was in full maturity during the Communist period. The Hitler years provoked painful memories of comrades lost in the war. Such moments would cause Edgar to break off from what he was saying, pause, and then—after recovering his voice—change the subject.

Erhard and Heidrun preferred to be interviewed separately. Like her father-in-law, Heidrun is a political activist and enjoyed the give-and-take of the interview. Erhard, more reserved, often strove to shape his response.

Their daughter, Maria, was given to talking about what was uppermost in her mind—namely, the mores of student life—but was not afraid to pass judgment on what displeased her. What Maria, her parents, and her grandparents all had in common was a willingness to share information about themselves. As time went on, they increasingly reflected on their past.

In all, forty interviews were held between August 1992 and October 1995, most of them open-ended. Toward the end of our work together, I prepared questions in advance to get information about particular subjects. On my return to Weimar, the tapes were transcribed.

Besides these formal sessions, we were invited to take lunch with Wally and Edgar on several occasions in their ample kitchen as well as to attend family celebrations or to join in guided tours by Edgar of the village and the Schorcht fields scattered around the outskirts of Göttern.

Apart from interviews, Edgar and Wally were generous in allowing us to make use of family documents. Especially important was *Der Bauer und sein Hof* (*The Peasant and His Holding*), containing detailed information on the operation of the Schorcht farm, compiled by Edgar in 1940 as a course assignment in agricultural school. Also useful were the legal documents pertaining to the transmission of the farm from one generation to another. In addition to these, the Schorchts gave me permission to make copies of numerous family photographs, with those of Edwin and Elly Schorcht, the principal protagonists in the first part of the book, holding special importance.

Archives

It was my good fortune that pertinent archival material was available. I made use of the Göttern chronicle—a survey of local events maintained year by year by the village chronicler—as well church records stored in the mother church in nearby Magdala, particularly useful when I was seeking out genealogical connections among the early Schorcht settlers in Göttern toward the end of the eighteenth century.

In Weimar, we consulted files in four archives: the *Stadtarchiv*, housed in the city hall in the *Marktplatz*; the *Thüringisches Hauptstaatsarchiv Weimar*, maintained in the former grand duke's stable (and later Gestapo headquarters) on Marstallstrasse, particularly useful for information on the land reform program of 1945–1950; the *Kreisarchiv Landkreis Weimar Land* on Schwanseestrasse, containing files on the relationship between the local administration of Göttern and district headquarters in Weimar; and the *Verwaltungsarchiv*, the administrative archive housed in the former Nazi Party headquarters in Thuringia. In addition, I made use of two archival sources lying beyond the confines of Göttern and Weimar: the branch of the *Bundesarchiv* in Berlin (formerly known as the Berlin Document Center under Allied Control) to amplify information given to me by the family in regard to membership in the Nazi Party, and the U.S. National Archives for data concerning the movement of American military forces in and around Göttern between April and June 1945.

Presentation

The book is divided into four parts, each reflective of generational rather than historical time.

Part I introduces the first generation of the narrative: Edwin Schorcht (who inherited the family farm in 1922 from his father, Hugo) and his wife, Elly (with whom he entered into marriage in 1920 in the early years of the Weimar Republic). The narrative deals with the historical derivation of the land, its cultivation, the material culture of the farmstead, and the location of the village of Göttern in space and time.

Part II centers on Edwin and Elly's children, Edgar and Irmgard; their coming of age in Hitler's Germany; their engagement with the war years and the later occupation of Göttern by the Americans and the Russians; their reaction to land reform and the industrialization of the countryside; and their coming to terms with the new Germany created by the USSR, leading to Irmgard's defection to the West and Edgar's emergence as the mayor of Göttern during the last half of the GDR.

Part III is the story of Erhard Schorcht and his wife, Heidrun; of their growing up in the Ulbricht years of the GDR; their matriculation at the University of Jena as favored offspring of workers and peasants, culminating in Heidrun's invitation to the United States in 1979; followed by her personal struggle with loyalty to the party in the light of her experience with American democracy.

Part IV turns to Maria, the daughter of Erhard and Heidrun. Also a child of the GDR, Maria finds herself looking back to the past in relation to the new Germany that history has bequeathed her.

Voices

The predominant voice in the following story is mine. When members of the Schorcht family speak, their statements, taken from the tapes, are indicated by quotation marks. On those occasions when I attribute behaviors or states of mind to persons, I am not presuming to read their thoughts or anticipate their actions but choosing to maintain the integrity of the narrative. To avoid imposing interpretive explication on the narrative, I have added a postscript to each chapter to include commentary.

PART I

1

Edwin and Elly

The Inheritance

EDWIN SCHORCHT took control of the family farm in the village of Göttern in 1922, on the occasion of its transmission to him by his widowed father, Hugo Schorcht. It was the fourth time that the holding had passed from father to son since the Schorchts had come into possession of the land at the end of the eighteenth century.

It was a day in late September when Hugo and Edwin drove the rig for the two-hour ride to the courthouse in Blankenhain, 14 kilometers away, accompanied by Edwin's three married sisters and their respective husbands. The act of transference, drawn up by a notary, was signed by everyone in the party; the duly notarized copies were then sent to the land registry office, the tax office, and the town hall.

The document, terse in its legal language but comprehensive in its intention, touched the lives of every family member present that day. It assigned to Edwin, in perpetuity, the entire 29.88 hectares of farm and woodland; all the livestock; the house, barn, and attached buildings; and all the movable goods and property, thus guaranteeing the transmission of the farm intact from one generation to the next.

With a stroke of the pen, Edwin had become one of the largest farmers in the village. In hand with his new status came responsibilities: toward his displaced father, his disinherited sisters, and, by extension, his brothers-in-law. For his father, Edwin was bound to guarantee a heated room on the first floor of the farmhouse, a place at the family table, spending money equal to the going market price of a metric hundredweight of rye in Weimar, the cost of medical treatment, and burial. And if for some reason Edwin should move elsewhere, the document specified that he must continue to make available to his father yearly ten hundredweight of potatoes, six hundredweight of corn, and a two-hundredweight pig.

For Hugo, the transfer of the farm promised a safe haven while simultaneously freeing him of the duties and responsibilities that had devolved upon him as its sole master at the time he assumed its control from his father, Wilhelm Schorcht, in 1893. His three daughters' liens on the farm, once their home, were of a different order than his own, less immediate but equally imperative. Meta, Liska, and Elsa had all married young farmers for whom the transmission of the Schorcht farm held promise for the welfare of their own holdings. The husbands' presence at the courthouse that day, where they affixed their signatures below those of their respective wives on the transfer agreement, bore testimony to their interests in its outcome, interests that were monetary rather than custodial. While the document spoke to the filial responsibilities of a son to a father, it just as clearly specified the financial obligations of Edwin to his sisters by requiring him to pay each of them 50,000 marks, the total being half of the appraised value of the property, with the initial 25,000 marks payable by January 1, 1923, and the remainder due three months after the death of their father. There were no liquid assets, however, by which Edwin could make good on his commitments.[1] In an attempt to lessen Edwin's burden, Liska and her husband, Hugo Schleevoigt, agreed to reduce the amount Edwin owed them by the sum of money they received from the sale of wood from one lot, a welcome gesture but hardly significant in the face of the enormity of Edwin's obligation.

There must have been a general understanding that day that the Schorcht farm would pass intact from Hugo to Edwin, even though each of Hugo's children had rights to an equal share of it, according to the prevailing laws of inheritance.[2] The tension between the imperatives of patrimony and those of inheritance was palpable throughout the social landscape of much of Germany, and most especially in the region of Thuringia, where Göttern was located. Until the twentieth century, Thuringia had been a patchwork of small sovereign states created by adherence to a doctrine of partible inheritance, which had successively sundered ducal domains almost as readily as family farms.[3] On this September day in 1922, the maintenance of the Schorcht farm was once again reassured through a balancing of rightful claims on its assets with a forbearance in favor of its integrity, thus escaping the fate of numerous properties that, at the crucial juncture of generational transference, were split up into small unprofitable plots by self-interested family members. The Schorchts, convinced of the greater benefits of keeping the farm intact, opted against division. When the family left the courthouse at Blankenhain, Edwin was the farm's new master, but at the cost of being saddled with onerous financial obligations and tense relations with his sisters and their spouses for years to come.[4]

Edwin: Student and Soldier

It was the widespread opinion in Göttern that Edwin Schorcht would be successful in whatever he undertook. Large-framed like all the Schorcht men, Edwin was exceptionally strong and energetic. A picture of him in his World War I uniform shows a tall, handsome man, all 6 feet 3 inches of him evidencing pride and determination.

Edwin was thirty-two years old when he finally stepped into his father's shoes, after working for him for more than ten years without benefit of compensation. In preparation for the day when he would be in charge, Edwin had enrolled in the agricultural school in Marksuhl near Eisenach, some 70 kilometers west of Göttern. There he learned the theory of agriculture, the most advanced technologies in crop cultivation, and the science of breeding stock to supplement the years of practical training he had gained working at his father's side before being called off to war.

He graduated in the spring of 1914, just months before being conscripted for military service. In September 1918, he returned to Göttern after serving in all four years of World War I. Trained in Leipzig in the Seventy-Seventh Field Artillery Regiment, he served in the Fifty-Fifth Artillery Regiment at Verdun. There he was wounded in March 1917, though not seriously. A letter from the front arrived in Göttern in April. His mother, guessing what news it held, collapsed when it arrived. She died of heart failure that same day. Hugo and his three daughters, with wife and mother dead and son and brother in daily peril of life and limb, consoled themselves as best they could with the thought that their unrelenting labor would keep the farm secure.

Verdun was an unforgettable experience for Edwin. A sentry and a spotter, Edwin was also responsible for helping bring supplies to advanced artillery positions. Every detail of the devastated countryside was burned into his memory. Twice in the years after World War II, he returned there, first with his daughter and later with other veterans on a bus tour, finding that he remembered the lay of the battlefields as if they were the back of his hand.

Edwin returned to Göttern in September 1918 wearing the Iron Cross 2nd Class. He immediately began helping his father set the farmstead to rights, as the war years and his absence had taken their toll. In his spare time, however, Edwin found himself taking notice of Elly Martha Maria, daughter of Louis and Thekla Michel, innkeepers in Göttern.

The Engagement

Edwin met Elly, four years his junior, at the inn where he went on occasion for a glass of wine. Edwin was an abstemious man, not one given to wasting time or socializing with friends and neighbors. Elly was quite a different sort. Pretty, she liked to sing and carry on a bit with her father's regulars, among whom were several who courted her favor. In the end, though, it was the man who went to the inn less often than the others who won her hand. Edwin drove to Jena to buy the gold wedding band by which they announced their engagement in 1919.

For several years previous, Elly, with her mother's help, had been slowly accumulating linens, with her maiden initials stitched in them, as well as other household goods and items of personal attire against the day she would marry. The large expense, though, once her marriage to Edwin was pending, was the purchase, in keeping with local custom, of a bedroom and living-room set. Louis Michel was able to save some on the final price by providing the cabinetmaker with wood.

The Wedding

The wedding took place on June 28, 1920. For several days before, friends and neighbors helped with the baking in large wood-fired ovens, while the father of the bride stocked the cellar with extra kegs of beer and bottles of wine. The evening before, the *Polterabend*—the traditional breaking of dishes—had already taken place in front of the inn. For days preceding, piles of old dishes had been collected in households throughout the village to make the greatest possible noise.

Photographs taken the day of the wedding show how fortunate they were with the weather. Family members began to arrive early by wagon, coach, and chaise from nearby Magdala and Milda, the birthplaces of Elly's father and mother, respectively; from Bucha, where Edwin's mother had been born; and from farther afield as well. With the prospect of continuation of the festivities well into the evening, it was no easy task to find stabling for the horses. Those who brought gifts—dishes and tableware—were given packages of cakes, each containing a small slice of all the different ones Elly and Thekla had baked in the preceding days. Some cake also went to villagers who, though good friends, were not invited to the ceremony.

The wedding began with the official signing at the registry, but it was the church service that was especially fine, with the flower girls spreading petals before Edwin and Elly as they moved slowly down the aisle, followed by Hugo, Louis, and Thekla, and then the more distant relatives. At the conclusion of the ceremony, Edwin was ready with a bounty of small change in his pocket to throw to the children as they crowded around when the wedding party left the church and proceeded to the inn, where Louis and Thekla worked but also lived.[5] The young men of the village had previously gone into the woods, returning with trees placed on either side of the door so as to construct an *Ehrenhof* (court of honor). The young women joined the trees together with garlands of pine branches. From the garlands was hung a sign that read, *Willkommen*. Before entering, Edwin and Elly stopped to partake of the *Brot und Salz* (bread and salt) offered them, representative of all that the couple would need in life. Tables had been set up both inside and outside the inn to accommodate the more than sixty relatives from the Schorcht and Michel families. First there was a clear soup with eggs; then the beef, pork, and veal schnitzel, along with dumplings, potatoes, and vegetables; to be followed by a "good dessert"; and, of course, all kinds of local wine: black currant, gooseberry, apple, hawthorn. Later there was dancing, an activity in which Elly took more pleasure than Edwin.

Edgar, Edwin and Elly's son, has in his possession a photo of the wedding couple and their guests assembled before the inn in Göttern. It catches a moment in their lives, each a farmer or a farmer's wife, dressed in their resplendent best as they put aside for a few hours the unremitting hard work that would be their daily lot. Another likeness is of Edwin in his good black suit, clutching a shimmering top hat, towering above his wife, also in black, evoking an image of tender solemnity.[6]

Finally, the gigs were hitched up for the ride home, so the cows could be milked and the horses and pigs fed, only to return again the following day, this time to Hugo's house for further pleasure at the table: coffee and Thuringian pastries of every variety to be consumed, toasts offered, pictures taken. It was customary to celebrate the first day at the bride's house and the second at the groom's, with the expenses shared accordingly. That evening, the finery was put away; the next day was work as usual.

Commentary

The trip to Blankenhain on September 28, 1922, was a defining moment for the Schorcht family, as was the wedding of Edwin and Elly two years earlier, on June 28, 1920. Both property settlement and marriage were fundamental to the production and reproduction of farm life in Göttern, Thuringia, and beyond. The forming of alliances between families forged by marriage and the consequent creation of another generation took place when property interests were transmitted from one generation to another. The effect of exclusion imposed by impartible inheritance on the aspirations of offspring created an atmosphere of tension and suspicion among siblings and between them and their parents, which was difficult, if not impossible, to allay. This was the case even when the younger generation appreciated the fact that pride in the family patrimony was contingent on the maintenance of its integrity.

It was the prospect of the acquisition of interests on one hand and the necessity of their renunciation on the other that provided the emotional context for the family members as they set out on their journey to Blankenhain. We can only surmise the feelings they harbored that day: Edwin's restrained exultation at the prospect of assuming control of what he had patiently waited for; the degrees of animosity and hurt experienced by Meta, Liska, and Elsa, unassuaged by the fact that they themselves had all become, like Elly, household mistresses in their own right, at the expense of their sisters-in-law; the remorse felt by Hugo at the relinquishment of his power, intermingled with relief at his surrender of responsibility for the welfare of the family patrimony; the unabashed anticipation mixed with suspicious regard felt by Edwin's three brothers-in-law (Hugo Schleevoigt to a lesser degree) at the prospect of being richer by 50,000 marks. We are probably right in assuming that none of these alleged emotions surfaced in untoward ways during the day, just as we may suppose that the passions generated by the transmission of the Schorcht farmstead were not transitory but left their mark for years to come.

We know this to be true in the case of Edwin. The responsibility, if not guilt, that Edwin felt with regard to his sisters and then later toward his daughter never left him. From the beginning, Edwin's strategy in operating the farm differed from that of his father in the emphasis he placed on breeding as the most effective means of raising the money to pay off the personal debts he incurred in fulfilling his obligations toward his sisters. Edwin's behavior extends the determining influence of inheritance emphasized by Jack Goody, Joan Thirsk, and E. P. Thompson beyond the realm of interpersonal relations to touch on, not surprisingly, economic courses of action.

By 1922, when Edwin assumed control of the farm, he already had Elly at his side. We have reason to believe that Edwin not only valued the proficiency that Elly brought to the management of the household but also cherished the support she gave him in confronting the costs—emotional and otherwise—that ownership of the Schorcht farm entailed.

2

The Lay of the Land

The Dispersed Holdings

ALMOST TWO YEARS PASSED before Hugo took the step that, while inevitable, could easily lessen a man's regard for himself. To pass on the farm to Edwin would require taking the backseat, becoming an object of charity in his own house. At age sixty-seven, Hugo was in reasonably good health, still capable of a day's hard work, but it was clear that the farm could benefit from Edwin's taking charge. He had proven himself at the agricultural school and knew better than his father how to deal with changes in cultivation and breeding. The years of Edwin's absence in school and in the war had taken their toll. Everywhere there was evidence of neglect. According to Edgar, Edwin's son, "Hugo was a good farmer, but he didn't work like *my* father. My father never stopped working. He said to me, 'Life is a struggle'; that is the way he lived." Hugo had the papers drawn up for the transfer of the 29.88 hectares of land that constituted the Schorcht family holdings.

The farm that Edwin inherited in September 1922 was divided into seventeen plots of arable fields and woodland as well as those on which the house, stalls, and barn were located. The holdings were widely dispersed throughout Göttern. No one in the village possessed a unified farm. Rather, each farmer's fields—some large, some small—were scattered, some in higher elevations, some in lower; some in richer soil, some in poorer. Edwin's largest field, the Waltersberge, the pride of the Schorchts, measured some 16.32 hectares and was situated on the opposite side of the village from the farm itself. The most fertile field, the Gommel—5.9 hectares lying directly to the east—could be reached more readily.[1] Three of his holdings were each slightly larger than a hectare; the remainder of his holdings were less. The smallest were arranged in long parallel strips interspersed among similar holdings of other farmers. That such a pattern of parcellation and scattering was an optimal agricultural arrangement cannot readily be argued. What is clearly apparent, however, is that the structure of the farm that Edwin inherited was a historical artifact, the product of a long process of political and economic contestation stretching back well before the Thirty Years' War.

The origin of the particular spatial arrangement of the Schorcht farm has partially to do with the system of agricultural production and the manner in which property was transmitted from one generation to another. Although interrelated, the way that farming was carried out and the tradition by which real property was inherited are historically independent of each other. Together they have left their mark on the German rural landscape.

The Three-Field System

It is in the open-field system of agriculture, and most particularly in the three-field (*Dreifelderwirtschaft*) variant of it, that the pattern of the dispersal of the Schorchts' fields finds its inception. First noted in western Europe in the early Middle Ages, the open-field regime was eventually to be found throughout the Great European Plain, reaching from the English Midlands to the Black Sea.[2] The widespread adoption of the three-field system in East Germany was brought about by the opening of these newly settled lands by colonizers from the West from the twelfth through the fourteenth centuries.[3] The open fields began to disappear rapidly in England with the inception of the enclosure of the commons in the mid–eighteenth century, whereas they continued unabated in East Germany until the middle of the nineteenth century and beyond.

Whereas Edwin's forebears in Göttern had firsthand familiarity with the communal nature of three-field agriculture, Edwin himself was to experience the reintroduction of communality, central to the collectivization of agriculture imposed by the German Democratic Republic (GDR) in the 1950s and 1960s. That both the earlier and later Schorchts knew agricultural communalism on the family farm in Göttern is not to suggest a genetic link between the two historical moments. The obligatory collectivization forced on Göttern by the GDR was a product of Soviet Russian history, whereas the communalism associated with the three-field system had its roots in the coordinated arrangements of cultivation and grazing inherent in the open-field agriculture as it existed in northern Europe for a millennium. Nevertheless, the reexperiencing of a communal agrarianism within the course of a hundred years by the Schorcht family, as by many others, remains a provocative fact.

It is because of both the role that communality has played in East German agriculture and the survival of the scattering of a man's land into the present that a discussion of the three-field system is appropriate.[4]

The three-field system had division as its defining element: the division of the village lands into arable and nonarable; the subdivision of the arable land into three very large fields; the further division of these fields into elongated strips, or *Gewanne*; and the subdivision, in turn, of the Gewanne into still narrower strips, representing the holdings of individual farmers. Thus, a person's farm consisted not of a consolidated plot of land but of a number of different strips scattered throughout the large fields.

At first glance, this fragmentation of farmland would seem to be perversely noneconomical, given the difficulties of maneuvering from site to site, as well as between the boundaries of one's neighbor's holdings. On further reflection, however, the survival of the three-field system over a period of close to a thousand years and the fierce resistance to its suppression through enclosure in the eighteenth and nineteenth centuries suggest that the scattering may well have proven economically functional, given the other features of the system, which have to do as much with integration as those just mentioned have to do with division.

Whereas the arable land was individualistic in its employment, each farmer sowing and weeding his particular strips, the nonarable land was communal in its utilization, used primarily for grazing the cattle and sheep of the villagers. Individual animals were organized into herds and grazed over a large area under the supervision of a herdsman. Cooperation of the whole village was involved in grazing.

In villages that were organized around manors, communal supervision was in the hands of the court; in nonmanorial villages, decision making was undertaken by a village council. Communal regulation also extended to the fallows—the one field in three that remained uncultivated, to be enriched by the droppings of the cattle. Communal decisions were also required in regard to not only the exigencies of large-scale grazing but also the cultivation of the arable land. Continuous cooperation was required for gaining access to individual strips, to prevent unwanted incursions, and to coordinate plowing and harvesting across individual strips.[5]

No matter how large any one individual's property might have been, not one of his strips, intermixed among others, was large enough to allow him to graze his cattle on it independent of the others. It is likely that the scattering of the arable strips helped safeguard the small cultivator's rights to communal grazing on the common and the fallow, lessening the possibility that the large owner might take advantage of the weaker. At the same time, it ensured that all cultivators took part in the collective grazing, decision making, and organizing of crops.

Although feudal obligations demanded labor service on the manorial estates and restricted a peasant's disposition of his holdings, the right to property transmission prevailed. Two strategies existed for this—impartible inheritance (*geschlossene Vererbung*) and partible inheritance (*Realteilung*)— and both were found in central Thuringia.[6] The eastern part of Thuringia, due to Slavic influence, is mostly characterized by *geschlossene Vererbung* (the property remains together in the hands of a single inheritor, with the others being compensated monetarily), and the western part by *Realteilung* (equal inheritance among all offspring).[7]

The Arrival of the Schorchts in Göttern

The first mention of the Schorchts in Göttern is a reference to Johann Heinrich Schorcht Sr., who moved there from Zöllnitz and married Anna Magdalena Stöckel in 1786. One year later, on March 11, 1787, he bought a farm and land from a member of the Stöckel family. But it was not until 1794, when his wife inherited considerable land from her father, who had divided his property between Anna Magdalena and his other daughter, Sophie Elisabeth, that the Schorchts become well established in Göttern.[8] The practice of partible inheritance over successive generations, as practiced in this case by Johann Anton Stöckel, led in time to the creation of plots so small they could barely be cultivated or plowed. Strips no more than 2 meters wide and holding fewer than 100 square meters were not rare. Loss through partible inheritance could be compensated for by purchasing additional land. The property acquired by Johann Heinrich Schorcht was located in the part of Göttern known as the *Anger* (village green), or the *Herrenseite*. This term dates back to the time when the *Herren* (notables) were the lords of Erfurt, whose lands in Göttern were administered by the office of Kapellendorf, to which the farmers of the *Anger* paid their dues.[9]

The First Transference of
Schorcht Property in Göttern

The first act of transference of Schorcht property in Göttern occurred in Kapellendorf on March 5, 1816. In this twenty-four-page inheritance contract, the forty-seven plots of land belonging to Johann Heinrich and Anna Magdalena Schorcht are listed, although no monetary value is assigned to them. The act transmits all the property to Johann Heinrich Schorcht Jr. (presumably the only son) under the condition that he take care of his mother's needs, her husband having predeceased her. It stipulates that he provide her yearly with room and board and provisions in the form of corn, wheat, barley, eggs, a fat pig, fodder for a cow, wood, and access to the gardens belonging to the property.[10]

The records show that in 1828, Johann Heinrich Schorcht Jr.'s wife, Sophie Magdalene Schorcht, purchased land in the Gauga district, the site of an abandoned village just south of Göttern.[11] Gauga, like two other small villages once situated between Göttern and Bucha, had died out, unable to keep pace with an emerging market economy.[12]

With a property that consisted of more than 25 hectares in 1846, the Schorcht farm was one of the more affluent in Göttern. In this position, Johann Schorcht was constrained to not only pay dues to three different princely domains, for each of which different portions of his land were fiefs, but also do socage for them. As had been the custom in previous centuries, his title as *Anspanner* obliged him, together with the other affluent farmers of the village, to do socage service with the help of his animals. In the case of the manor of Magdala, this meant ninety-eight days of work on the fields per year. Less-privileged farmers, *Hintersiedler*, were required to perform manual labor.[13] Even so, their subjugation to manorial obligations did not impose a loss of personal freedom.[14]

Relief from these feudal duties was set in motion by the reforms following the revolutionary fervor that swept over Germany in the middle of the nineteenth century. As the records show, the farmers' protest in Göttern and their partial refusal to do socage for a number of years resulted in their partial removal in May 1846. For their freedom, the farmers of Göttern were obligated to pay 700 Thaler to the manor in Magdala. The respective documents show the signatures of all the farmers concerned, the first on the list being that of Johann Heinrich Schorcht, *Schultheiss* (mayor)—a position his great-great-grandson was to occupy more than a hundred years later.[15]

The farmers' dissatisfaction was fuel for the 1848 revolution, after which a decisive law regarding the removal of all manorial duties was passed on April 14, 1848. Feudal taxes could be paid with the sum of twenty-two times the yearly fee. The government opened state bank institutes (Ablösungskassen, Rentenbanken, Kreditbanken) to offer loans to help during this process of indemnification.[16]

Records show that on September 14, 1855, Johann Heinrich Schorcht Jr. paid off his feudal dues once and for all for that quarter of his land that had been a fief of the dukes of Sachsen-Gotha-Altenburg. Additional payments to the grand duke of Sachsen-Weimar-Eisenach and to the governor (*Statthalter*) of Erfurt, as the representative of the elector and archbishop of Mainz, ensured that by 1860, the Schorcht farm had become free from all feudal obligations.[17] But the splitting of the land placed severe limitations on the ability of an individual to succeed in effectively managing a farm in the burgeoning market economy. Accordingly, the agrarian reforms involved not only paying off feudal fees but also rationalizing agriculture through the privatization of the commons and the consolidation of the strips of land.

A major impetus for these changes came, in the case of Göttern, from the liberal administration of the grand duchy of Sachsen-Weimar-Eisenach, determined to convince the farmers of the value of a constitutional monarchy and thus avert recurrence of the revolutionary moment of 1848 by establishing a firm base of farmer proprietors.[18]

In 1858 in the Kreis Weimar, in whose jurisdiction Göttern lies, 8,664 people owned 172,120 pieces of land, or an average of twenty pieces of land per owner. The consolidation (*Umlegung*) of the land decreased the number of pieces by one-fifth, and in some areas by one-quarter. The structure of ownership was not fundamentally changed by the consolidation, although middle-size and large owners were strengthened as small owners weakened.[19] The consolidation was perceived as a mixed blessing by some, as James Sheehan points out: "The consolidation of fields into separate units made innovation easier, but also removed an important source of community solidarity."[20] The theme struck here by Sheehan, of regrets for a passing solidarity following the demise of a rural feudalistic structure in East Germany with the introduction of a full market economy, would be heard again with the abandonment of collectivization in 1989–1990.

Resistance to the transformation of the three-field system was particularly strong among the small and marginal farmers, who saw that they had the most to lose in the passing of the traditional mode of communal grazing. The abolition of the rights to use the commons, central to the reforms, was the source of the resistance. Peasants were for the most part against consolidation. Because accessible pastures for their few cattle were always scarce, they depended for their survival on the right to graze their livestock on fallow fields and the common lands.[21]

In Göttern, the consolidation of the fields took place from 1862 to 1873, in accordance with a law passed on April 23, 1862. Every farmer was attributed land equal in quality to that which he had owned before and subsequently had to relinquish.[22] To compensate for the variety of soils, the consolidated farms were of different sizes, so that each had a representation of different soil qualities. In addition, paths had to be built through and around the former open fields, so that each farmer could now have access to his individual fields. In 1847, the Schorchts' property consisted of eighty-one separate holdings. Following the consolidation, it was reduced to twenty.[23] In all, it was a costly undertaking, involving detailed surveying, lengthy calculations, and a multitude of negotiations to bring it to completion in eleven years. As the chief accountant (*Rechnungsführer*), Wilhelm Schorcht, Edwin's grandfather, was responsible for paying the bills incurred in

Göttern's consolidation.[24] Today Edgar Schorcht, Edwin's son, thinking about that phase of the family's history, wonders out loud why Wilhelm, considering his position, could not have done better for the family, allotting it more of the favored Gommel field and less of the Waltersberge, which is so difficult to work with due to its incline and stony soil.

The Schorchts as Independent Farmers

By 1870, the Schorcht farm had assumed the form it would still have in 1922, when Edwin took control of it. Liberated from feudal service obligations, fees, and the communal responsibilities of the three fields in which the family holdings had been subsumed, Wilhelm was free to run the farm, now made more manageable by consolidation, growing whatever and whenever he wanted. Jerome Blum, writing about the process of consolidation, notes that "metamorphism was the single most important departure from traditional agriculture, heralding as it did the transition from communalism, with its collective rights and controls, to individualism, with its private rights of property and its individual freedom of action."[25] The way was open to apply the innovations in cultivation and agricultural technology that were moving apace in the nineteenth century.

Some improvements had already begun to make their appearance toward the end of the old order, chief among them the adaptation of the "improved" three-field system, or "green fallowing"—the planting of fodder crops, especially lucerne and clover, on the formerly bare fallows. Planting forage freed the farmer from some of the constraints of a traditional agriculture, for it allowed him to raise more and better livestock. More livestock meant more droppings and thus the possibility of growing new crops that required heavy applications of manure. In addition, more forage meant more stall feeding and thus the freeing up of former pasture land for plowing.[26]

New crops were also brought in, chiefly the "Americans"—maize and potatoes. At first there was strong resistance to the potato, perceived as suitable for only animals. As Sheehan writes, "The turnabout came during the years of the grain crop failures and famine that engulfed Switzerland, Germany, and Austria in the early 1770s. Hunger swept away the peasant's scorn for the vegetable."[27]

The change from an extensive system of cultivation to an intensive one was facilitated through the use of chemical fertilizers, particularly useful in the expansion of sugar-beet production.[28] These root crops not only played a crucial role in the new forms of crop rotation but also provided a link between industry and cultivation, being used as raw materials for such industrial products as starch, alcohol, and sugar.[29]

Credit became available through banks and through cooperatives.

In nearby Magdala, a Raiffeisen Bank was opened. Friederich Wilhelm Raiffeisen founded, in 1849, the first of a series of cooperatives designed to provide farmers with credit and investment capital.[30]

After the crop failures in the 1840s, the next three decades were a time of noteworthy prosperity for many German farmers. Theodor von der Goltz in his *Geschichte der deutschen Landwirtschaft*, first published in 1903, writes of the years between 1859 and 1880 as "the happiest period that German agriculture had ever experienced."[31] Farm prices increased across the board after mid-century, with rents and land values following close behind. The Schorchts took advantage of this fortunate period to strengthen and maintain their position as one of the leading farm families in Göttern. They prided themselves on being among the first to plant the new fodder crops, thus enhancing the quality of their livestock. Edgar remembers his father talking about the earlier Schorchts and their forward-looking ways when it came to agriculture.

Wilhelm and Hugo had witnessed the fundamental changes in the mid–nineteenth century that led agriculture out of the subsistence economy of the Middle Ages toward association with an emerging world-market economy. Now freed from the pervasive communal restraints that characterized traditional German agriculture from the early Middle Ages, encapsulated as it had been in the manorial system, they strove hard to make good on their stewardship of the newly configured Schorcht farm. They were among the first generation of independent eastern German farmers. They were to exist as such for a scant ninety years before the collectivization program of the GDR was to reintroduce agricultural communalism, although in a different form, in the late 1950s and early 1960s.

Hugo's son Edwin was born in 1889, at a time when German agriculture entered a period of crisis as foreign competition began to drive down agricultural prices and land values. Russia and the United States started to export grain and meat. "As a result, Germany lost the French and English markets, and its cattle exports to western European countries were reduced."[32] Consequently, many German farmers were in a precarious economic position in 1890. Strengthened by the consolidation, the Schorchts weathered the difficult years and then gained from the higher prices for agricultural goods brought about by the agrarian tariff policy, first initiated by Otto von Bismarck and then continued, although in modified form, by his successors: Leo von Caprivi; Chlodwig, Prince of Hohenlohe-Schillingsfürst; and Bernhard von Bülow.[33]

From Wilhelm to Hugo

Wilhelm relinquished control of the farm to his son Hugo on April 15, 1893. In the contract, Hugo acknowledges that his parents have the right of free and general access to all the property attached to the farmhouse as well as use of the middle living room and the bedroom standing off of it, with both rooms properly heated and lit. In addition, they are to receive free meals at the same table as Hugo's family (*Kost am ungesonderten Tisch*). Hugo agrees to pay them pocket money of 4 marks a week and money with which to buy clothes (50 marks a year). In addition, Hugo is obligated to pay his brother, the farmer Edwin Schorcht of Kleinschwabhausen, the sum of 15,000 marks to indemnify him for his share of the inheritance. Both brothers agree to raise a gravestone (*Denkmal auf dem Gottesacker*) for their parents at the time of their death.[34] Wilhelm and Adelheid would die within a month of each other in 1903.

Hugo was thirty-eight years old, well into manhood, when he acquired the family property from his father. He had married Alma Becher from Bucha in 1888 in the same church where years later his grandson Edgar would take Wally Huenneger for his wife. Hugo, it is said, was not the farmer that his father had been nor the one his son would become, but he made more than ends meet as he and others capitalized on the soaring food prices that prevailed in the pre–World War I period. Especially successful were the returns on the traditional crops that Hugo sowed—wheat, rye, potatoes, and sugar beets—cultivated and harvested with cheap farm help from Poland, whom Alma lodged and fed in quarters across the courtyard above the washroom. At the same time, greater productivity was achieved by the introduction of sowing, threshing, and grain-mowing machines. Furthermore, Hugo, aware of the gains to be made from livestock following the decline of sheep herding in Thuringia in the latter part of the nineteenth century, began to breed cattle, making use of the newly introduced Simmentaler stock.[35]

The War and Its Aftermath

Following the declaration of war in August 1914, the optimism that had prevailed among farmers gave way to discontent as the government imposed price controls on agricultural products in an attempt to provide adequate food supplies at reasonable prices for the troops at the front and for the workers in war-related industries. The creation of the War Grain Office in 1915 and the War Food Office in 1916 failed to stem the decline in the output of agricultural goods, as farmers withheld produce despite pleas for their patriotic collaboration, seeking instead outlets for their crops and animals in a flourishing black market.[36] Hugo found himself working harder for less, as the availability of cheap Polish farm labor dried up, as did the supply of fertilizer and replacement parts for machinery. In addition, farmers like Hugo who before 1914 had enjoyed a sense of privilege and protection now began to experience a relative decline in their social and political statuses vis-à-vis industrial workers.[37]

Although thoroughly embittered by the agrarian policies of the kaiser's government, farmers by and large remained loyal to the government and played no part in the widespread labor militancy as the war drew to an end.[38] It was not surprising that farmers from Göttern, disaffected as they were, were not present in Weimar on November 9, 1918, when members of the Weimar Worker's Council met with representatives of the Soldier's Council to demand the abdication of Grand Duke Ernst of Sachsen-Weimar-Eisenach.[39]

Nor did the government of the republic, established by the Constitutional Congress in Weimar in 1919, win the support of the farmers, most especially as the new republic felt constrained to continue the controls made necessary by the worsening situation in the countryside in the months following the cessation of hostilities.[40] By 1920 and early 1921, however, persistent resistance to fixed prices and delivery quotas had made the agrarian controls invalid—except for those pertaining to prices for grain, the commodity that affected Hugo most directly. The economic and political instability that followed the initiation of a controlled economy was further exacerbated by the deliberate policy of inflation that early Weimar governments undertook to pay reparations to the Allies by printing paper money. By the summer of 1922, German currency began to experience uncontrolled deterioration, prompting acute shortages, hoarding, and a reinforced antipathy of the agrarian class toward the Weimar Republic.[41]

It was against this background of uncertainty, ambiguity, and widespread hostility toward the Weimar Republic that Hugo's son Edwin took control of the Schorcht farm on September 1, 1922. A man of considerable rectitude, he must have felt grateful at the prospect of putting things to right with the farm following the unsettled years of the recent past. We can, in fact, only surmise what he may have felt on that day; what we know for certain is what he inherited from his father on that day: a farm that has a story to tell, revealing connections between the past and the present.

Commentary

What is striking about the above account is how relatively late the capitalization of agriculture occurred in central Germany, as opposed, for example, to that in England and France. It was still possible, if not necessary, for a farmer to be involved in feudal agricultural practices in much of Thuringia through the first four decades of the nineteenth century.[42] The myriad social and economic reasons for such belatedness go beyond the more modest intent of this commentary. Instead, I would like to focus on those processes that farmers, such as the Schorchts, undertook by necessity to become independent. In so doing, I am suggesting that the steps taken to capitalize agriculture in the middle of the nineteenth century in the feudally organized grand duchy of Saxony-Weimar-Eisenach—namely, the consolidation of fields—are comparable to those undertaken to decapitalize it in the middle of the twentieth century in the socialistically organized GDR.

A crucial task faced by farmers in connection with the enclosure of the common land had to do with altering the relations of production, meaning that farmers had to liberate themselves from those feudal relationships that curtailed their ability to act entrepreneurially on their own behalf. The abolition of feudal rights was one of the consequences of the revolutionary moment of 1848, which brought liberal reforms in its path. As revolutionary struggles go, these were relatively easily won. Their success was aided by the landed proprietors, who saw gains for themselves in the redistribution of former common lands. So, too, the commercial classes regarded changes that would enhance activity in the growing market economy as serving their own interests. Thus, when Johann Wilhelm Schorcht set foot on the stage of private entrepreneurship circa 1860, he was in fine company, albeit overdue in his appearance in comparison with his western European counterparts. Actually, he may not have made his cue in 1860, for apart from winning control of the means of production, he had as yet to rationalize them.

The economy of scale operating in the field system—as well as the need to arrive at communal decisions in regard to when to plow, when to fertilize, and when to harvest—would serve to inhibit attempts to cultivate profitably in response to market demands. It stands to reason that the difficult and painful task of consolidating the strips into more tractable fields was undertaken communally; the decision to consolidate was done village by village, with Göttern beginning in 1862. The actual work of calculating how a farmer's plots were to be added to here, diminished there, and consolidated with those of another, while keeping in mind the major criteria for restructuring the plots—quality of soil (in a village like Göttern, soil quality varied markedly between hillsides and valley), distance from farmstead, and topography (workability)—was the responsibility of the surveyor, and all of this had to be done as farmers continued to plant, cultivate, and harvest their crops. One can only imagine at what emotional cost these transactions occurred—the wills that clashed, the patience tried— before the consolidation was completed in Göttern in 1873.

The cultivatable land left available following the consolidation was, in fact, somewhat smaller than what had existed before; roads had to be built to make each field accessible, and hedgerows and field markers had to be put in place. To cover these costs as well as those for the initial surveying, farmers took out bank credits. Given the quickening economy that prevailed in the last quarter of the nineteenth century, repayment occurred relatively quickly.

When Johann Wilhelm experienced metamorphosis from feudal dependent to individual farmer, he recapitulated the transformation that feudalism underwent to capitalism at the same time as the Schorcht property became adapted to capitalist agriculture through consolidation. Ironically, consolidation on a larger scale was called on a hundred years later to readapt the same fields to meet the demands of collective agriculture, as the great-grandson of Heinrich Wilhelm underwent his transformation from a capitalist to a socialist farmer.

3

Farming

Daily Life

EDWIN, THIRTY-THREE YEARS OLD, was well prepared when he took over the farm in 1922. He had gained from his father's experience by working at his side. In addition, unlike Hugo, he had been formally trained at agricultural school. Edwin undoubtedly believed, too, that he had acted wisely by entering into marriage with Elly Michel. Although the daughter of an innkeeper, Elly had been brought up in Göttern and knew farming ways. She was, according to her children, a small woman but physically capable. In time, she became an accomplished milker, able to milk their nine cows in less time than it took Edwin. Elly took as evidence of her prowess the large head of froth she was able to produce in the buckets she filled.

As far as the farm work was concerned, every day was the same: Edwin and Elly rose between 4:30 and 5:00 to feed the animals, clean their stalls, and milk them. Edwin, knowing the particulars of bovine rumination, realized that feeding and milking should not be hurried. As his son Edgar put it, "He understood that if you didn't put in the time, you didn't get anything back." After the milking and feeding, the milk cans were cleaned in the washroom, and the fresh milk was readied to be picked up for delivery to Jena by 6:00. Then it was time for the first cup of coffee before going off to get feed for the animals.

The Schorchts had only three-quarters of an acre of suitable meadowland, not enough pasture to allow all of his animals to graze outside. Edwin was aware of what a toll it took on a cow's lungs—indeed, its whole body—to go without exercise day after day, and there was always the risk of pneumonia.

Nonetheless, he was forced to feed the milking stock in their stalls. The cows' enforced quiescence was a source of pain for him, but there was nothing he could do about it. He allowed the young stock to graze outdoors so that their growing bodies had a chance to develop properly and thus withstand the stress to be placed on them later. Edgar says, "Our situation was not like that up in Schleswig-Holstein, where the cows are out in the fields all day and their birthing process is easier. We had cows that stood all day—that were fed and raised just to produce milk. The birthing process was always a big risk. When a neighbor's cows had a hard time in birthing, people would run and get my father, and he would pull it out. That was his reputation in the village."

After caring for the animals, Edwin and Elly would stop off in the kitchen on their way to the fields for malt coffee. Like other Thuringian farmers, Edwin and Elly preferred cake made of oats, milk, and butter to bread at that hour; the cake was eaten with stewed fruit, usually prunes, cooked in a large vat and stored in stone pots. There was no time in those days to make marmalade or other delicacies.

It was then that Hugo would join them. In the beginning, he worked as hard as the others, giving a hand in the stables, but as time went on he cut back, eventually restricting himself to helping with the hay or weeding.

Shortly before 7:30, Edwin would see to hitching the team of horses to the wagon to get to the fields as soon as possible. If the Gommel was their destination—the choice 5.9 hectares located on the eastern flank of a gentle rise behind the farm buildings—then it would suffice to move out through the barn and be at their destination in a matter of minutes. But if the task at hand necessitated going to the 16-hectare field, the Waltersberge, then they would need to leave more time to pass through the village and slowly mount the 1,000-meter hill that rose majestically from the town cemetery to the crest of the Jena forest that shaded the village to the east. On days when manure was to be spread on the Waltersberge, they would make a number of trips. On these days, Edwin saw to it that the horses received an extra ration of oats, above their usual allotment of 5 to 7 pounds. Later, when Edwin raised and kept Belgian horses, he found that for the same amount of work, a Belgian needed at least 20 pounds of oats daily.

From the top of the Waltersberge, one could look across the valley of the Magdel to where the Dierzholze, a smaller Schorcht holding, lay on the other side of the Autobahn. Indeed, today, if one knows where to look, the other twelve pieces of fields and woods belonging to the Schorchts, dispersed about the township of Göttern, are visible from that vantage point.

When the wagon was unloaded on the upper slope of the Waltersberge, where "it was nicer," the horses were let free to graze at will, if they were not needed in particular areas to "keep the grass short." If it was plowing time, then Edwin would set to with the team. Working well, he could expect to finish 1 *Morgen* (0.25 hectare) in a day, although he needed more time on the Waltersberge, given its pitch and the number of stones in the soil.

By 9:00, it was time for breakfast: more malt coffee served with bread, wurst, pickles, and apples. The coffee was brought up in a clay jug, wrapped in a towel to keep it warm, and laid in a basket of straw. It was a moment to rest the body and, if the earth was warm, to stretch out and relax. Edgar remembers those occasions from his boyhood: "Those were always the nicest of times." But Edwin never allowed the moment to continue for too long; there was work to be done.

In those days, the sowing of wheat was favored. It did well in Göttern soil and, indeed, in all of Thuringia. Rye was also grown in abundance. Although its straw was longer than that of wheat and not so well suited for animal bedding, it was useful for making rope. In the early 1920s, the cutting was still done with a scythe, and the threshing by beating. With two, three, and four strokes, the kernels were beaten out and the straw softened so that it could be readily made into rope: fifty strands of straw per shank of rope. Edwin had learned that skill at agricultural school.

In addition to wheat and rye, oats were grown (for the horses) as well as beer barley, sold to the breweries in Ehringsdorf or Apolda. Then there were the labor-intensive crops—potatoes, turnips, and onions—which had to be planted, dug, and cleaned by hand. The walk home for the noonday meal was not begrudged. It was a break in the routine of work, an occasion to stretch out tired limbs and be given food. Edwin led the return from the fields at 11:30 sharp. Before eating, the horses had to be fed, and if the cows had recently calved, they had to be milked again, as their udders would be full.

In the kitchen, food was on the table: pickles and *Schmaltz* (fat drippings) to spread on the dark bread. Some found cooking at noon to be too much work, but Elly saw to it that a warm dish was always at hand, if only dumplings or rice gruel (the latter was apt to provoke a comment from Edwin)., Edwin and Elly's daughter, would later note that her mother always got "worked up" when her father complained about a meal, saying, "When it's good, you don't say a thing; when it's bad, you complain." Three days a week there was meat: poultry, pork, or beef. A cow was slaughtered once a year and the meat preserved in jars. Irmgard remembers how sensitive her father was. Sometimes the meat in the jars would start to turn, and Edwin, annoyed, would chide Elly and ask in a loud voice, "Do you smell that? What's that smell?"

There was always a good number at the table for the noonday meal, even though the immediate family was small. Edgar, Edwin and Elly's first child, was born in 1922. Irmgard followed in 1923. Room at the table was made for them in due time. When Elly's mother, Thekla, died in 1927, her father, Louis Michel, joined Edgar's father, Hugo, for meals as well.

The hired help—two men who worked with the animals and two women who worked in the kitchen as well as in the fields and stables—also ate with the family. Before the war, one of the men was Walter Zimmermann, who was married and lived with his wife and children in the village. Walter would be at the farm the whole day, working with the horses. The other hired man lived on the farm and helped take care of the cows and the pigs. The two "maids" came mostly from the Thuringian Forest area, where poverty was pervasive. Because they were known for their fine singing voices, Elly would encourage singing during housecleaning, and she would even join in. Edgar remembers how the whole house would resound with their voices on Saturday mornings. The help received board, meals, and wages—7 reichsmark (RM) a week—which was 1 to 2 RM more than was given at any other farm in Göttern. In those days, you could buy a good pair of Manchester trousers (corduroys) for 7 RM. The other farmers would sourly ask Edwin, "How can you pay them so much?" Edwin, getting good results, was satisfied he was doing the right thing.

During planting and harvesting time, there were always a few women from the village around the kitchen table as well, who helped out in the fields. Rarely were there fewer than twelve people for the noonday meal. If everyone sat down to eat at noon, Edwin was on his feet again by 12:45, ready to return to the fields.

In the afternoon at harvest time, when the sun had lain on the fields for hours, wheat was cut and bound in sheaves, five or six of which would be stacked together. After a week, when thoroughly dried, they were brought to the barn in the wagon. It was also the time to bring in the forage crop of rape, once it was no longer damp. Sometimes it was known to happen that a sack or two of root crops were overlooked in the field in the gloaming. At mid-afternoon, malt coffee and cake were again brought out to wherever the work was, signaling a time to stop, sit down, and stretch aching limbs.

Depending on the time of the year, the work in the fields stopped at 6:00, or possibly 5:00, to leave enough time for the hour and a half necessary for milking and feeding. The stalls had been illuminated by electricity since the turn of the century, but Edgar can remember petroleum lamps lit by pinewood spills from when he was a child. He would be sent on an errand to the Gaststätte, where his grandfather Louis kept a 200-liter petroleum barrel next to the door, operated by a small pump. He also recalls when there was no running water in the barn, even though the first tap had been installed in the kitchen in 1903. The water was pumped from the well behind the barn into a large 150-gallon tank suspended in the stalls, and it flowed from there into the water troughs, keeping them at best barely half full. Only later, when running water was introduced in the stalls, was it possible to regulate the water level with an automatic shut-off mechanism.

The evening meal, *Abendbrot*, was set on the table by 7:00 or sometimes as late as 7:30 if the chores ran over. There was wurst, cheese, *Schmaltz* to be spread on bread, and potato soup. Edwin and the men in general liked soup in the evenings, especially if it was cold outside. Irmgard remembers her mother saying to them, "You say you want 'just soup,' but a good soup is a lot of work if it is going to taste good."

Elly continued to work in the stables as well as prepare the meals. At first she cooked over a peat-fired stove. Edgar remembers, when he was a young man, driving the team to Hohenfelden, 20 kilometers away, for a wagonload of peat. Later, wood from their own lots was used until brown coal was introduced. It seemed that Elly had as much of a flair for cooking as she did for farm work. Edwin, who was not given to excessive compliments, praised his wife's potato soup.

Hugo joined the others at the table for dinner, his presence there guaranteed by contract.

After dinner, Edwin read the paper, while Elly tended to her handiwork. Once a week, water was heated for bathing in the big vat in the washhouse. Everybody knew that taking too many baths was not good for you; it depleted the skin of essential oils. Before going to bed, recourse was made to the toilet in the courtyard, a visitation not happily undertaken in the cold. The waste flowed into a large container, which, when full, was driven out to the fields, "the best kind of fertilizer," it was said. By 9:00, everyone had retired to bed.

Later, Edwin was proud to be the first person in Göttern to have a cast-iron tub installed in his house, the hot water for which was heated by a coal-fired oven. But that was not until 1937.

Sundays

On Sundays, work had less to do with productivity than with presentation. While the women prepared the special Sunday meal, the men spent the morning in the stalls. The horses were thoroughly cleaned with the currycomb until their coats glistened, their hooves blackened, their harnesses polished. After the cows were brushed clean, their tails were washed with soap and water. Edgar says, "We were obsessed with our animals. On Sundays, the women would ask, 'Do you even take a toothbrush with you to clean their teeth?' It was our hobby, really."

With so much to do to get everything in order, there was no time for the family to go to church together. One or more persons would be designated to go for the others. It was usually the elderly, or if someone were about to have a confirmation, he or she would go too.

The noonday meal inevitably included Thuringian potato dumplings (*Klösse*) and *Braten* (roastmeat) pork or beef, but more likely the former than the latter. As Edgar says, "There are families who equate dumplings with Sunday. We would have potatoes and a vegetable, but of course, we had our dumplings as well."

The noonday meal was extended until 2:00 or 3:00 in the afternoon, depending on the weather. If a thunderstorm or rain was foreseeable, Edwin would wish to get out into the fields as soon as possible to bring in more of the harvest. In the afternoon, there were always things in the stable to do; in the summer, there was work in the fields. The garden also needed attention. There was little time for leisure.

Holidays

The most important holidays were Christmas and Easter, which were observed with family members. On Christmas Eve, there was a spruce tree, children opened presents, and carols were sung; carp was the usual meal. Easter focused on the church service, but both holidays involved a good deal of extra baking and cleaning.

Ascension Day, the fortieth day after Easter, and *Kirmes*, which in Göttern is celebrated on the last Sunday in October, were the two most important holidays after Christmas and Easter. Extra slaughtering would take place to take care of the friends and neighbors who would drop by. On the Sunday before, the stables were thoroughly cleaned, as were the animals, so you could bring in your guests and say, "Here, have a look at this":

> Wally: They always wanted to see everything.
> Edgar: That's just the way it is when you have guests.
> Wally: Otherwise you wouldn't do it.
> Edgar: So, you are forced to do things.
> Wally: We cleaned from the basement to the roof.
> Edgar: All the windows.
> Wally: Well, we still clean them now and then!

As for organizations, there were not many in a small village like Göttern. There was no *Schützenverein* (shooting club), as there was in many German communities. According to Edgar and Wally, it was too expensive; no one could afford it.

Edgar remembers the mixed choir that Herr Venus, the schoolteacher, organized. "That was really great," Edgar recalls. He also recollects his mother's sewing circle (*Spinnstube*), which met once a week during the winter months. The ten or so women would take turns meeting in one another's homes. The serving of bean coffee on these occasions was a special treat. Edgar says, "Then with the Nazis it slowed down, became more political. It was still possible in the Nazi time, but they wouldn't meet in the same way. It was called the Women's Organization (*Frauenschaft*). The center was in Magdala. They would go down there once a month to meet in the Gaststätte. But it wasn't the same thing." Edgar continues, "Then in the GDR [German Democratic Republic], it got a second wind, the DFD [Demokratischer Frauenbund Deutschlands], the German women's organization. The Götteners took an interest in it and developed it. It wasn't the same in every village." Wally says, "They met once a week! Young and old. Now it is only the retired women."

Baking

For most of her childhood, Irmgard remembers helping her mother bake the cakes that were eaten first thing in the morning and the bread consumed later in the day. Baking took place once a week. The freshly ground flour that Edwin brought back from the mill was kept in a large wooden box with three compartments: one for rye flour, another for wheat flour, and the third for fine white flour for special occasions. The rye flour for bread was sieved the night before so that it could acclimate to room temperature. The cakes were made first, eight to ten of them: large round cakes, cakes made from wheat, fruit cakes, and crumb cakes to sate the appetite of the large household.

The sourdough for bread was stored in blue-gray pots, with a little salt sprinkled on top. Every other week some of the old sourdough was used for the new batch. The dough was kneaded, sprinkled with water and salt, and left overnight to rise. In the morning it was kneaded again, shaped into small loaves and set in a row, and left to rise again. Later it was kneaded a third time and molded to fit into "pans" of straw and left to rise. Elly cut a cross in each loaf before placing it in the ovens.

The two ovens, too long for the kitchen, were recessed deep into the wall of the washroom across the courtyard from the kitchen. Irmgard and her mother cut to size the bundle of sticks the men brought down from the woods, bound them, and left them to dry on the roof of the potato silo. After these were burned in the ovens, the bottoms of the ovens were wiped clean of the ashes. First the cakes were baked, for they needed the most heat. The remaining heat was sufficient to bake the bread, the loaves being withdrawn by a long wooden paddle before they were finished and sprinkled with water to give them a glazed appearance.

Irmgard also remembers the hours spent mending, washing, and ironing clothes. The soap was made from pig's fat. A small piglet was slaughtered and cooked in a big pot, left to cool, and then cut up to extract the fat. It took three days to do all the washing for the household, scrubbing the hand-sewn work clothes clean in a large wooden tub made by the cooper. After drying, the wash was cold-pressed with a hand-operated press. The good shirts and dresses were ironed. At first a heavy iron filled with coals was used. One had to be ever vigilant lest an ember fall out and burn a hole. Every four weeks, when the washing was done, an elderly woman came out from Magdala to help with the mending. Elly mended as well and made new white and pleated shirts for Edwin, Hugo, and Louis. They had to be snow white when they were clean.

Breeding

In 1928, Edwin decided to add another string to the farm's bow: animal breeding. He would start with bulls and add horses. It was not that his father had not paid heed to breeding; in fact, there was a time when Leitholf and Schorcht bulls alone serviced the eighty cows of Göttern. Walter Leitholf and Hugo Schorcht had formed a breeder's society (*Zuchtverein*) and had their stock registered. But while Hugo had relied on traditional methods of breeding, Edwin wanted to bring the full weight of what he had learned in agricultural school to bear on the improvement of his stock. There was another reason, too—the payment to his sisters of the settlement for their rights to the farm. A farmer, following good breeding practices, could expect to set aside a substantial sum.

The first step was to add to and strengthen the bullpen in the northeast corner of the courtyard. Given their size and temperament, bulls were always a possible source of danger. Edgar says, "The bull stall had to be stable. [*Laughs.*] When the bulls were three to four years old, they were the size of small elephants! They were dangerous beasts. Sometimes at night we would be awakened, and the bulls were loose in the barnyard! We would have to run out and get them."

When it came time to buy stock, Edwin traveled south to Franconia, to towns like Ansbach, where he was assured of getting the best buy for his money. In the early 1930s, a healthy animal could be purchased there for 450 RM, while a breeding bull would fetch 900 RM, although some sold for as high as 1,000 to 3,000 RM. After a close inspection of the bulls at hand with due regard to their registration papers, Edwin would arrange for the transportation of the animal in a freight car north to Grossschwabhausen, the nearest rail head to Göttern, and he would ride in the car with the bull. From Grossschwabhausen, it was a 7-kilometer walk by road, albeit a bit shorter through the woods to the farm, with Edwin leading the animal by a piece of rope attached to its nose ring.

Using the bulls to mate with their stock as well as other cows in Göttern was a practice that Edwin had already learned about from working with his father. He was confident, however, that he could now make better-informed judgments in matching cows with certain qualities with bulls that would complement these characteristics. Usually the promise bore fruit; when it did not, the animal was sent to the slaughterhouse. In time, Edwin's judgment in these matters brought an increase in stud fees and the knowledge that by raising the quality of the animals in the community, he had won for himself a reputation as one of the best breeders in the region.

Raising bulls for sale was a new venture for the Schorcht farm but one near and dear to Edwin's heart. Having become a member of the Cattle Breeders Association, Edwin was invited to participate in the bull auction in the large wooden hall on Riesnerstrasse in Weimar. His hired man, whom Edwin had taken on to help with the animals, would clean the bull, and when his turn came, Edwin would lead the bull to be auctioned around the ring. The longer the bull was led, the higher the price went. Up to 1,000 RM, every raising of a hand was worth 20 RM, and more than 1,000 RM, every raising of a hand was worth 50 RM. The highest price paid for a bull during Edwin's time auctioning was 7,000 RM, but that came after several years of careful breeding. Edgar says, "We raised and sold at least three bulls a year. If we didn't move them at the auction, then we would get the butcher and he would take them away. We didn't take them back home. If we sold them as breeders, we would get at least 900 marks. If it was a bull with good blood, it would go up to 2,000 to 3,000 marks. And a 1,000-mark note was worth something back then."

Over the years, Edwin learned from experience that everything depended on the parents. A small defect could be overlooked if the father and mother had a good bloodline. According to Edgar, Edwin believed that if you did right by your bulls, cows, horses, and pigs, then you would do well. As Edgar puts it, thinking about his father, "A breeder has to be a perfectionist; he has to put everything into it."

One measure of Edwin's success in breeding was the milk production of his own cows. A milk inspection was made each month by the animal breeder bureau. An inspector examined the cows in the mornings and the evenings and determined how many liters each cow produced as well as the fat content of the milk. A record book was kept for each animal. Among the five farmers in Göttern who maintained cow herds, the Schorchts, on an overall basis, performed the best. In the early years, if a cow was giving 2,000 to 3,000 liters, she was considered a good producer, but in later years, just before the farm became collectivized in 1960, Edwin had doubled the production through good breeding practices, obtaining results of 4,000 to 5,000 liters a year as well as a significant increase in the fat content. At first, Edwin was able to fill four 20-liter cans a day. Later, with good breeding and more and better feed, he could fill six. More and better feed, however, meant straining to increase the fodder production on the Schorcht fields, land that was not ideally suited for that purpose.

Horse Breeding

With bull breeding proving to be profitable, Edwin turned to heavy Belgian draft horses (*Schwere Belgier*). Breeding horses required making alterations to the existing stall space off the courtyard. To start, Edwin acquired at auction a stallion and a "mother-stud." The mother-studs, when foaled, were either sold or bred again, depending on the market. Of the two foals delivered each year, Edwin would raise one. After three years, it would be large enough to start working. At first, this horse would be used only as a "third horse" to run with the others when they were hauling the wagon and get accustomed to the noises. Edgar says, "With us, it wasn't like it was in Texas; we trained them gradually."

Edwin preferred the Belgians because of their strength, which they needed for hauling the manure up the "mountains." Going down the hill required good brakes, because even the Belgians could not hold the wagon without them. For working on the flat, Edwin liked the Oldenburger breed. They were fast but not so strong.

Edwin had no difficulty selling the Belgians, but he started breeding them too late to make a financial success of them as he had with the bulls. Before the war, tractors were beginning to make their appearance in Göttern farms, anticipating the demise of the horse. Edwin and other farmers resented the way the heavy tractor compacted the chalky soil of Göttern, making it difficult to work.

Edwin continued to use his horses even after he bought a tractor of his own in 1941. He enjoyed working with the horses more than he did driving a machine. Wheeling about a pair of heavy Belgians at the end of plowing a strip never ceased to be a pleasure. His favorite horse was a magnificent stallion named Felix. Even after Edwin retired and was well into his seventies, he continued to take care of Felix. Whoever needed a field plowed could borrow the stallion.

Bulls went the way of horses when artificial insemination was introduced after World War II. An insemination station was created in Erfurt in the early 1950s that used better, more expensive bulls, costing as much as 10,000 RM each, which Edwin could not afford to buy himself. As a result, Edwin began to purchase semen from the station for his cows. Göttern was one of the first towns to make use of artificial insemination in the district. As Edgar says with pride, "Wherever the newest thing was, my father was there too."

Commentary

At the heart of daily life on the farm was routine, the repetition of the same tasks day in and day out. Only measured changes were allowable— inadvertent ones were intolerable. Just as it was a moral injunction for their masters, predictability was regarded as efficacious for the animals. Cows that were milked on schedule, pigs that were fed regularly, horses that were worked or exercised daily—all produced, performed, and grew optimally. Edwin was regarded as a disciple of routinization as well as cleanliness: stalls regularly whitewashed, animal hides curried, tails washed, hoofs blackened. Order prevailed around the courtyard just as it did in the house under Elly's supervision, the weekly schedule of baking, cleaning, and washing intermeshed with the daily task of food preparation. Elly was well regarded for her organizational skills, essential for not only the fulfillment of the household tasks but also the management of the necessities of those nonfamily members who lived and worked with them: the hired hands who prevailed before the war, the prisoners of war who predominated during the hostilities, and the apprentices of the postwar period. For Elly, it was a sacred trust that the noonday meal should appear on the table at twelve o'clock. As her daughter, Irmgard, now says, "My mother was a very punctual woman. My father was pleased with that side of her. It all had to work out timewise."

The two domains—the market one of the barnyard, stable, and fields, and the domestic one of the household—existed in a synchronized and complementary relationship with each other. Both spheres were essential for the success of the farming enterprise as a whole, even though Edwin's prerogatives in the place where he dwelled were not equaled by Elly's in the barnyard and stables. It was Edwin's right to enjoy the comforts and orderliness of his home and to pass judgment on what was set before him at mealtimes, without at the same time having anything to do with the work necessary to accomplish the results he expected. On the other hand, there were liens on Elly's time and energy (or on those of another woman paid to work in her stead) above and beyond those expended in the home that required her to perform certain gender-specific tasks in the productive sphere of the stable and field.

Elly's relationship with the farm's livestock, however, was distinct from that of Edwin. A woman had responsibilities toward cows that she did not have for prestigious horses, bulls, or oxen. It was her task to milk the herd in the morning as well as those cows that required it at the end of the workday. Milking demands a notable expenditure of muscle power in the hands, wrists, and arms as well as in the legs in the repetitive motion of getting up and sitting down. The feeding of the cows, however, unlike the feeding of their masters, was the prerogative of men. It was Edwin's daily undertaking to select and mix the feed in the feed stall in proportions that had been effectively tested in the agricultural school at Marksuhl. Elly had also learned about ingredients and their proper proportions at the cooking school in Jena she had attended when she was sixteen. But no one—including Elly—would attribute the same importance to frequenting a cooking school as frequenting an agricultural one, not because the welfare of animals surpasses that of humans but because one is the purview of men and the other of women. So it was with baking and breeding. Elly baked large quantities of cake and bread every week of her adult life, constituting an essential part of the daily diet of those working on the farm. Edwin supervised the breeding of his bulls and cows far less frequently. Nevertheless, the Schorcht farm became renowned for its breeding; Elly's efforts were taken for granted.

What is of interest here is not only the gender-specific valorization that was imposed within the context of farmstead work on the Schorcht farm in the years preceding World War II but also the differential quality of the work that was expected of the two sexes. In the field as well, some of the most toilsome if not menial tasks were assigned to women: cultivating rows of growing plants, digging potatoes and beets, clearing fields of stones, spreading manure.

We have no way of knowing how Elly and Edwin would assess the equity of the distribution of the kinds of work done by men and women during the years they were managing the farm. It may very well have been that they viewed the sex roles of their time as appropriate, if not natural. We are privy to the opinions of Edgar and Wally, however, who believe that on the whole, a disproportionate degree of burdensome tasks fell on the shoulders of women.

If we have reason to believe that this assessment is generalizable beyond the confines of the Schorcht farm, it would seem that it was something akin to a process of colonization, menializing rural women, imposing on them an inequitable share of work, the results of which men took credit for. This turn away from familial communality of effort, in which there was agreement on routine and procedure, toward gender inequality in the workplace is a dynamic of rural life worthy of more notice than it has received. In light of this understanding, further discussion of the place of work in the daily life of a farm like that of the Schorchts would move from family to gender. Family is an analytic point of departure, not an unassailable given. It is this perspective that Hans Medick and David Sabean are speaking to when they say, "In this way of viewing the matter, family and household are no longer taken for granted as fixed magnitudes. One looks through and beyond family to consider changes in sexual relationships, questions of inner and extra-familial sexuality, childhood, youth, old age, and death."[1]

4

The Farmstead

The Courtyard

IKE THE OTHER FARMSTEADS in Göttern, the Schorcht house, barn, stalls, and sheds are situated in the village, at some remove from the fields surrounding the settled area. This distancing of dwelling from field finds its historical source in the three-field system, in which cultivated land was scattered around a central cluster of dwellings.[1]

For the newcomer to Göttern, and indeed to Thuringia, the first sight of what is referred to as a "central German, closed, three-sided farmstead" (the prototype of the Schorcht dwelling) is intriguing.[2] From the outside, one appears to be looking at a redoubt, the only access to which is a door set into the large gate, the gate joining two sides of the three-sided structure. Only by opening the door does one gain access to what lies within: a courtyard enclosed by the dwelling and stalls on one side; additional stables and stalls on another; and to the rear, the barn, on the third side.

The courtyard (*Hof*) is the central space in the integrated whole, which is the farmstead (*Gehoeft*). From there, one has immediate access to the milk room, the stalls and sties, the hay loft, the hen coop, the rabbit hutch, the feed bins, the manure pile, equipment, and the house situated around it. With so much that one needs close at hand, steps are saved in the early morning that are later expended in moving from field to field; the historical disbursement of property in the open fields stands in inverse relationship to its rationalized integration in the barnyard. On more than one occasion, Edwin altered the arrangement of animal stalls to enhance efficiency, keeping in mind the sequence of steps required for their care and maintenance.

The Schorcht courtyard is some 30 meters long and almost 6 meters wide. It is widest at the rear, whereas at the front, at the foot of the gate, it narrows to 4 meters. As barnyards go, it is not large; indeed, it is the narrowest yard along the *Anger*. Its small size was a matter of considerable irritation to Edwin, all the more so for a breeder requiring, on occasion, space for his animals beyond that afforded by the stable, to say nothing of having to maneuver hayricks and other large farm equipment.

Affixed to the front of the house, just to the right of the gate, is the small but recognizable blue-and-white logo of the Bureau for the Preservation of Historical Monuments (*Denkmalpflege*). The monument status of the Schorcht farmstead has been awarded not because of age but for the traditional arrangement of house, farm structures, courtyard, and gate. Changes to any part of it are not allowed without permission from the proper authorities.

The prototype of the central German *Gehoeft*, of which the Schorcht farmstead is an example, achieved its present form in the late eighteenth and early nineteenth centuries, a time of increasing rural wealth and concomitant concern about the protection of property. The farmstead of so-called angled farm construction, enclosed on multiple sides, secured a farmer's person and that of his family and goods against marauders.[3] Once inside the courtyard, an intruder, if not immediately set upon by a mastiff, could only wonder from which window or aperture he was being observed and would soon be apprehended.

Threat of fire constituted another danger, especially for interconnected structures, such as three- or four-sided farmsteads. It is not surprising to hear that Johann Wolfgang von Goethe himself took a personal interest in the hazard of fire to farmers in the Weimar countryside, prompting the head commissioner of construction in the dukedom, C. W. Coudray—an eminent architect in his own right—to design more fireproof farms.[4] The fact that the Göttern fire of 1833 was as devastating as it was, given the abundance of water flowing in the Magdel close at hand, was undoubtedly due to the ease with which it spread from one dwelling to another.

In a four-sided farmstead (*Vierseithof*), space, accessibility, and surveillance were maximized. Such "ideal" structures were more common in elongated street villages (*Strassendörfer*), where more space was available between adjacent farmsteads. In more circumscribed villages, such as those built around a village green (*Angerdorf*), the four-sided arrangement could best be achieved through a strategic marriage between neighbors or the purchase of an adjacent property.[5] That the Schorchts did not attain such an ideal says less about their ambitions or aptitudes than it does about the self-regulating mechanisms of a village like Göttern, where three-sided farmsteads are the norm.

The House

Access to the courtyard and thus to the house has varied from one historical moment to another and, in so doing, passed comment on the perceived security of the time. In the olden days, the *Hof* door was left unlocked. Neighbors could enter the courtyard and knock on the kitchen window. During World War II and afterward, when the Russians were around, it was locked, but during most of the German Democratic Republic (GDR) era, it was left unlocked. Now, since unification, it is left locked much of the time.

The existence of the *Hof* allows its owners the privilege of privacy beyond the confines of the house, allowing one to pass from the interior domestic space to the external enclosed courtyard before entering the public world of the street. The *Hof* provides an intermediate environment between the private inner and the public outer sphere, analogous to the Mediterranean courtyard in so far as both provide protection from public assessment, but different from it as well, in that the *Hof* is specifically familial.

When Edwin took over the farm, the ground floor of the Schorcht house consisted of two sitting rooms in the gable end, facing the street; the family room (*kleine Stube*); and the parlor, or "good" living room (*gute Stube*). The family room was narrow, with a sofa, several chairs, and large ceramic stove. Sitting on the sofa gave one, with a minimal amount of craning of the neck, a privileged view of what went on in the street, permitting a careful assessment of each passerby without being seen.

A room for company, the *gute Stube* was 50 percent wider than the *kleine* (which it adjoins), with two windows looking out onto the street, as well as two onto the courtyard. Here the best furniture was kept, and on the wall hung landscapes. In the glass-doored side cabinet was the special china, carefully arranged, including a set of Weimar porcelain, made in Blankenhain and initialed "ES," purchased for Edwin and Elly Schorcht at the time of their wedding by Edwin's mother and father. Against the north wall stood a majestic, tiled, dark-green stove. Generally, the *gute Stube* was closed off; it was heated only on holidays or family days: baptisms, confirmations, weddings and funerals, or those Sundays when coffee and cake were served in there to friends and neighbors.

The remainder of the ground floor beyond the two sitting rooms consisted principally of the kitchen and storeroom. The entrance to the house and the stairs to the first floor were located between the kitchen and the adjoining cow stalls. In 1946, Edwin, with the help of his son, Edgar, who was recently released from the army and a carpenter, undertook major changes on the ground floor. For some time, Edwin and Elly had been contemplating alterations. With Edgar home from the war, it seemed appropriate to build for the future. A wall was removed so that the kitchen, always deemed cramped and dark by Elly, was joined to what had inadequately served as a room for eating; this larger space proved more than ample to hold everyone for noonday meals. The large airy room, with windows looking out onto the courtyard, became the center of the household. With the kitchen in place, a more spacious entrance to the house was fashioned between the sitting rooms and the new kitchen, with a staircase leading to the first floor, where the four bedrooms, each leading from a hall, were located.

The Animals' Quarters

Adjoining the house along the north side of the courtyard was the cow stall, some 25 feet long and 20 feet wide. In 1923, the year after he had taken over control of the farm, Edwin set about remaking, with the help of a mason, the old half-timbered (*Fachwerk*) stall with brick.[6] With the cattle being stable-bound throughout the year, and thus more liable to infection, Edwin wished to provide them with the most durable housing he could.

At first the cows, all *Simmentaler Höhenfleckvieh* (Swiss spotted), were lined up in a row, as Hugo always had them, even though Edwin was not convinced that this was the best arrangement. It was not until 1937, however, that he finally decided on a radical change: placing them head to head, five on one side and four on the other. With the feed trough running down the middle between them, he found that they ate better and produced more milk. Edwin liked the *Simmentaler Höhenfleckvieh* because of their high-fat-content yield: 4 percent, depending on the fodder, while others would not exceed 3.3 percent.

Adjoining the cow stable was the feed stall, where a large variety of forage was kept for the animals, as dry and free from termites as was possible: lucerne, clover, and soy for the cattle, as well as a sour feed mixture of rye and vetch; oats for the horses; potatoes and sugar beets for the swine; and corn and bran for the chickens.

Next to the feed stall were the bulls. In any one year, Edwin maintained two breeding bulls and as many as four younger bulls, raised to be sold. He could have sold more than that, but the courtyard was already too small without usurping more of it for further stall space.

As for feed, the bulls were given essentially the same thing as the cows, with the addition of a greater amount of oats, which provided more protein for their sperm. Breeding took place in the courtyard. A cow was ready for mating when the cow "thread" appeared at its rear. She would then be led into a depression in a rear corner of the farmyard, hollowed out for the purpose, allowing the bull to stand somewhat higher than his mate. She was held fast in the front, with one person on either side to the rear, but if she was willing, there would be no problem once the bull was brought out. As there were always two bulls, the lads took turns.

The bull stall abutted the barn at right angles, closing off the courtyard to the west. Across the yard from the bulls was the stall for the younger stock, adjoining the barn on the south side of the courtyard. Here the calves were kept after being taken from their mothers' udders, as were the heifers being raised for market.

Adjoining the stall for younger stock was the pigpen, capable of holding up to forty pigs. The piglets were kept here with the sows as well as the fattening stock and runners—a runner (*Läufer*) being a piglet fed separately from the mother. Particular attention was paid to the porkers to make sure they received the proper ration of potatoes, sugar beet, barley, and skimmed milk to fatten them up for slaughter. Those showing the most favorable characteristics were raised as breeders. When a sow was ready, the boar would be brought over from the neighbor's farm.

For family consumption, at least two pigs were slaughtered and dressed in the spring (Ascension Day) and again in the fall (*Kirmes*). That figure grew as the number around the table increased, as it did after World War II, when refugees crowded around the table with the Schorchts.

On the day of the slaughtering, the courtyard was a busy place, as Edwin, Edgar, and the hired hands worked the animal's legs to pump out its 6 to 7 liters of blood to be made into blood sausage. The carcass was then dunked in a large tub of hot water with a bit of tar added, by which much of the animal's hair would come off by itself. The remainder was removed with a scraper. When hung by its hind legs, the carcass needed to be carefully dressed with well-sharpened knives. Scores of glass jars had to be washed for preserving the meat seasoned with salt, pepper, and cumin, to be made into different kinds of wurst. During GDR times, the skin was consigned along with the meat to make up each farmer's quota. As for the piglets, they were sold in the market, except for one or two required for the making of soap. One of Edgar's earliest memories is being bundled up by his mother for the long carriage ride to Jena with a large basket of squealing, squirming piglets at his feet.

Along the outer wall of the pig stall and extending for 2 meters into the courtyard was the manure pile, enclosed on its three sides by high boarding. Twice a day the pile got larger, as mucking from the stables was added, and six times a year it was depleted, as the manure was carried out to the fields. "If the barnyard had only been bigger," as Edwin lamented, they would not have had to spread manure so often. Getting the heavily loaded wagons up the hill was a strain for the horses. Three wagons were used in relay: one that was filled, one along the way, and a third being emptied on the hill. It would take ten to fifteen wagonloads to get three narrow rows of manure up the 16-hectare Waltersberge hill. The men worked on the wagon, while the women spread the manure, the hardest work. Edgar says, "Then the field had to be plowed right away; you can't leave manure lying out in the open air too long: it will let go the ammonia and everything else. . . . It's a real science. That's why we went to agricultural school."

From the summit of the manure pile, a rooster summoned all to wakefulness in the early morning. In the summer, swarms of flies hovered and buzzed over the pile, while hens pecked at its base.

Adjoining the pigpen on the south side of the courtyard was the horse stall. For a while it had been on the other side, where the bull stall was, as Edwin experimented with the optimal location for it. In 1925, he rebuilt the horse stalls in brick in a fashion similar to what he had undertaken for the cows in 1923. The stallion and the two mares were kept in the horse stalls, while the foals were housed with the other young stock. Edwin felt a special fondness for his horses. His well-muscled Belgians were not only splendid to look at; they also were his companions as he plowed, following behind their sure and steady pace. To own horses, to be a "horse farmer" in a village like Göttern, was an unassailable sign of distinction.

The last structure on the south side of the courtyard was the washroom, where the milk pails and cans were washed in the morning and again at night after the second milking. In 1927, Edwin rebuilt the existing washing facilities in brick, with a cement floor to facilitate cleaning. Washing milk pails and cans was a chore assigned to children, once their parents were assured they could do it properly. It was a job no one coveted, as getting wet was unavoidable. When clean, the pails and cans were set upside-down on a wooden rack in the courtyard to dry. The hot water for washing was heated by a large boiler that was also used for washing clothes. Here, too, were the ovens for baking, recessed into the wall of the washroom, one set above the other. Edgar can remember from his youth how on baking days, the aroma of freshly baked cake and bread wafted through the *Hof,* enticing all in its vicinity.

Between the washroom and the horse stall, a steep and narrow staircase ascended to a first floor with three bedrooms. For years it served as lodging for retired parents. Louis Michel, Elly's father, moved in after Elly's mother, Thekla, died. During the war, Polish farm help slept in the rooms, as refugees did later. In April 1945, American troops occupying Göttern commandeered the Schorcht house, forcing the family to move across the courtyard. Through the years, the use of these rooms has reflected the changing destinies of the family.

The Barn

When Edwin took control of the farm, the barn was already crowded. Hugo had had the impressive structure built in 1912 to Wilhelminian proportions.

Even so, its amplitude barely sufficed. The mower-baler (*Selbstbindermäher*) that Hugo had purchased in 1912 took up considerable space. Even bulkier was the Zimmermann baler (*Zimmermannstrohpresse*), which Hugo had acquired in 1921. Edgar recalls that the baler was always in the way when the rig was driven into the barn, requiring the heavy piece of machinery to be laboriously moved. Modern machinery was larger and heavier than what had existed when the *Hof* and the original barn were laid down in the preceding century.

The Massey mower, an American-made left-sided cutter, required three horses to pull it. The mower cut the grain and tied it in bundles. The bundles, left to dry in the fields, were later carted to the barn and stored there until the winter, when an electric-powered thresher did its job. Before electricity was installed, energy was provided by horses walking in a circle while harnessed to a center wheel, turning it and transmitting power to the thresher through a series of belt wheels.

The 1930s, which saw the Nazi regime "aligning" (*Gleichschaltung*) the farming population through its "blood and soil" policy, allowed the Schorchts to purchase additional equipment with money acquired through subsidies. In 1941, Edwin purchased a Lanz Bulldog tractor, manufactured in Mannheim, without the benefit of subsidies. Edwin paid the handsome sum of 5,400 reichsmarks (RM) for the Bulldog, knowing all too well what the introduction of tractors would mean for the eventual fate of his beloved horses.

The Bulldog's 20-horsepower motor was not powerful enough to allow Edwin to plow straight up an inclined field like the Waltersberge, as would a modern 100- to 200-horsepower tractor, forcing him to plow it horizontally. Despite its limited capacity, using the Bulldog for the heavier plowing, harvesting, and hauling took some of the burden off the horses, leaving them to be employed for the corners and the precision work. When it rained hard, chains were placed over the rubber tires of the tractor to provide more traction in the softened earth. If necessary, the rubber tires could be replaced with iron wheels.

Starting the Bulldog, especially on a cold morning, required both perseverance and heat applied to the front cowling. When cranking, precautions had to be taken against the dangers of backlash from the large flywheel. Once started, the Bulldog was left idling in the field, its distinctive down-cycle thump audible at a distance.

Edgar remembers driving bulls to auction behind the Bulldog as far as Gotha, some 60 kilometers away on the new Autobahn, largely bereft of traffic. The tractor's top speed was 20 miles an hour, so the trip consumed three hours each way—an experience Edgar enlivened by taking the tractor out of gear, allowing bulls and all to roll down each hill.

Behind the barn were three cement silos—two for dry fodder and one for steamed potatoes—all built in the early 1930s at prices made affordable by the Third Reich. The silos were fed by machinery owned by the Raiffeisen Cooperative. The Schorchts used the straw cutter, as did the Lydolphs and the Müllers, "large peasants" (*Grossbauern*), like them. The cutter was needed only in the fall, when the corn and sunflower harvests were brought to the silos. The cutter threw chaff directly into the silos, with the whole operation taking far less time than had been the case with the hand-operated cutter.

For potato steaming, Raiffeisen formed a steam cooperative (*Dampfgenossenschaft*) for Göttern and nearby Magdala and Ottstedt. Edwin was among the first members. The large steamer, fueled by coal, contained three kettles linked to the steam chamber by hoses, capable of steaming 600 kilograms of potatoes every half hour. From the kettles, the potatoes were emptied into a wagon, from which they were bagged and carted to the silos. There the bags were deposited under a layer of dirt, providing feed for pigs for the entire year. Steamed feed keeps its nutritional value longer; furthermore, pigs fed on steamed, rather than raw or even cooked, potatoes fattened up more quickly. Edwin found that the pigs were always sated after eating steamed potatoes, whereas before, when given uncooked potatoes, they would continue to squeal long after mealtime.

The Status of the Schorcht Farm
on the Eve of World War II

By 1939, Edwin had been at the helm of the Schorcht farm for seventeen years. He had worked hard, enjoining others to do the same. Evidence of success was apparent at every hand. He had become one of the most successful animal breeders in the region and, in doing so, had increased the farm's milk production. He had used the knowledge that he had acquired in agricultural school to improve field cultivation through a variegation of crops and to introduce new methods of planting and harvesting as well as fertilization. Valuable property near at hand had been purchased and an additional barn built on it, allowing Edwin to better handle the increased productive capacity of the farm.

All of this was achieved in the face of certain obstacles, one being the chronic lack of adequate help. Both Edgar and Irmgard were growing into good farmhands, but they hardly sufficed. Additional help was always necessary, requiring feeding and lodging as well as remuneration.

A further concern had to do with the chalky nature of the soil on the Schorcht farm. Chalk, while fertile, is hard to work. Some of the fields had what Edwin called "five-minute soil," meaning you either finished plowing it in five minutes or you spent hours trying to till it. If it hardened, you had to use rollers that broke the soil to aerate it and then firmed it down again. Optimal soil was rated 100. On that scale, Schorcht land had an average of 40, with the treasured Gommel ranging between 60 and 70. According to Edwin's children, these deficits were challenges for their father to overcome and, in doing so, proof of his worth as a farmer.

Farm Income and Expenditures for 1939

In the course of his last year at the agricultural school in Weimar, Edgar was required to compile a booklet titled *Der Bauer und sein Hof* (*The Peasant and His Holding*), on the state of the family farm for the year 1939.

The total income was calculated at 12,697.50 RM. The largest single source was derived from the sale of milk, sold in the milk market in Jena for 19 Reichsfpennig per liter. Crops ran a close second, with wheat being the largest earner, followed by rye, barley, oats, potatoes, peas, and straw. The Raiffeisen Cooperative in Magdala bought the wheat, rye, barley, oats, and straw, whereas a dealer purchased the potatoes and peas. In 1939, Edwin sold four young bulls in Weimar at auction, one nonproducing cow to the central slaughterhouse, ten piglets to a retailer in Jena, and 1,250 eggs to a dealer.

On the debit side, the largest sum paid out in 1939 was for wages, including for artisans: a blacksmith for shoeing, a saddler to repair harnesses, a cartwright to fashion a new wagon wheel, a cooper to repair wooden barrels, and a fitter to work up a metal coupling for the mower. The second-largest expense was for chemical fertilizer purchased from Raiffeisen to supplement manure, followed by cow feed and seed.

According to Edgar, Edwin felt his most onerous costs to be taxes: property tax, poll tax, dog tax, slaughtering tax, and church tax.[7] There were also insurance premiums for fire, hail, and animal losses, including death from epidemics, and coverage for general liability.

Domestic expenditures included the cost of brown coal, used in the tile stove in the family sitting room (the only room in the house besides the kitchen that was heated); electricity; use of the radio; the chimney sweep; Edwin's newspapers; and Elly's thread.

Income exceeded expenses, leaving a handsome surplus of 5,172.55 RM. According to Edgar, Edwin intended to set aside approximately 1,500 RM for family use, including clothes for Irmgard. Irmgard was being noticed by the young men of Göttern, although Elly was of the conviction that none of them were good enough for her. Irmgard was of that age when she always wanted to look her best. She was, in Edwin's opinion, much like her mother in terms of having expensive tastes: a dress here and a hat there.

Edwin himself was not loath to buy himself a bit of new apparel when the fancy hit him. Tall and dignified, he could have been mistaken for a man of affairs. Nor was he reluctant to spend a little money on travel. Edgar remembers going with his parents and Irmgard to the Rhine and Bavaria once during the "Hitler years" on a trip sponsored by the Reich Food Estate. Edgar says, "My mother liked to drink a little wine and be gay, while my father would climb the highest mountain in Berchtesgaden. My mother told him, 'Go on up; I'll be down here watching you and drinking a little wine.'" Now, though, with a war in the making, Edwin would forgo any excursions. When better days returned, there would be time to stretch his legs in Bavaria. For the present, he would put aside 2,265 RM for depreciation costs and as a starter for the year ahead, 490 of which would be earmarked for new equipment. For his father, Hugo, he would allocate 500 RM. Edwin was still obligated by the contract of 1922 to provide the old man with pocket money; the 500 RM would come out to about 1.70 RM a day, enough for a beer or two at the Gaststätte. He would make sure that Edgar assumed the same responsibility toward *his* parents when the time came.

In 1939, though, Edwin had more immediate concerns. With Edgar about to be called up in the army, one can only imagine Edwin's mixture of pride and apprehension, similar to what his father must have experienced when Edwin went off to war in 1914.

Commentary

The family narrative in this chapter has largely had to do with the status of the Schorcht farmstead for the nineteen years from 1922, when Edwin took it over, until 1941, when he purchased his first tractor, nine years after the Nazi Party came to power. This period of Edwin's tutelage is one of improvement in the general welfare of the farm—especially for the years 1933 to 1941. The balance sheet for 1939 showing the farm well in the black testifies to the farm's good fortunes, as do the refurbishing of equipment at subsidized prices and the addition of innovative facilities in the form of the silos for fodder storage installed in 1933 and 1934 behind the barn. Edgar's and Wally's perceptions confirm this impression:

Edgar: Those weren't hard times.
Wally: The farmers were supported then.
Edgar: To prove his power, Hitler had to get rid of unemployment. He built the Autobahn. When you were working again, then you had some bread and money. There was also a new inheritance law [the Reich Entailed Farm Law; see Chapter 6]. When we bought new machines, we received subsidies, as we did for the silos in the

back, which were new for the times, as was the layered manure pile. There were powerful subsidies for all that stuff. Things went well for farmers then, yes, for all farmers. In the beginning, they didn't ask too many questions about politics.

Edgar's reflections on how "things went well for farmers," especially during the early Hitler years, does not correspond to the general critical assessment of the state of agriculture during the Nazi dictatorship. The major problems facing German agriculture—inadequate production, manpower shortage, decline of agriculture in the overall German economy—were not solved and, if anything, worsened during the dictatorship of the Third Reich.[8] These conclusions are not to be disputed, but they do regard aggregates of information that are collated at the national level, whereas the data in Edgar's farmer's book are specific to the Schorcht farm. When Wally says that "farmers were supported then," she is testifying to the effectiveness of the National Socialist German Workers' Party's (NSDAP's) propaganda in making the German farmer feel, at least initially, to be the "cornerstone of the German State." At the same time, Edwin was able to keep ahead of the labor shortage by using Polish prisoners of war and by raising production. Part of the disparity between the recognition of the failure of the Third Reich's agrarian policy and the Schorchts' economic success during those years lies in the realization that understandings arrived at the level of macrohistory are not necessarily confirmed by what is known to have taken place at the level of local history.

PART II

5

Edgar and Irmgard

Birth

EDGAR SCHORCHT, Edwin and Elly's first child and only son, was born on January 25, 1922. In those days, it was customary for children to be delivered at home, with a midwife from Magdala in attendance. It was only later, after World War II, that mothers would normally give birth in a hospital in Weimar or Jena.

Edwin later recounted that Edgar was blue when he was born, with the cord wrapped around his neck. He said, "The midwife picked him up by the leg, unwrapped the cord, and gave him a hard smack on the back that got him breathing again."

Irmgard, a baby daughter, was born a year later, "on the third day of Pentecost," Irmgard likes to point out. Following Irmgard's arrival, Elly had several miscarriages. Edwin and Elly would have no more children. Consequently, it would be necessary to rely all the more on hired help to work the farm in the years to come.

Following the deliveries of both of her children, Elly spent most of the required week of rest in bed or in her room. The midwife dropped in during the day to weigh the babies and attend to Elly. As soon as she could, Elly got up and about, and in a matter of weeks she was back to her regular chores. When she wasn't feeding or holding them, Elly would place the babies in the wash basket. Playpens came a good deal later.

"Max and Moritz"

As they grew up, Edgar and Irmgard were inseparable. "Max and Moritz," as they were called, referring to the widely read children's story by Wilhelm Busch, would set out, pails in hand, busying themselves with games and pranks first in the farm's courtyard, then later in the village and in the fields and woods beyond.[1] There was no end of things to do and to explore. In the winter, they would skate when the pond froze over and, when it snowed, would slide on the Gommel hill behind their house. Their cleverest scheme was to hitch the two goats, Moritz and Hans, to the sled so that they could ride up the hill in comfort. In his later years, Hans grew too fat to be of much help.

By the time the children were seven and eight, they ceased going about together. Irmgard began making friends with other girls her age in the village, inviting them to her house or going to theirs. She and Anna Rausch became best friends. For Edgar, there was only one boy in the village born in 1922, Eric Fechte, the others being vintage 1921. The two boys spent endless hours playing cops and robbers until darkness drove them home. "I must say we had an unsupervised childhood," comments Edgar with a chuckle, thinking back on those days.

The fact of the matter was that Edwin and Elly, busy all day with the operation of the farm, had little or no time to be with their children, other than at mealtime or bedtime. Fortunate for both Edgar and Irmgard was the presence of their mother's father, Louis, a friendly, knowledgeable man.

When his wife's health began to decline, Louis sold the inn in the village where Edwin and Elly's wedding feast had been held in 1920 and bought a house in Weimar on Brehmestrasse. After his wife died in 1927, Louis spent most of his time with his daughter and son-in-law in Göttern, occupying a room across the courtyard above the washroom, which he shared with his grandson during most of Edgar's childhood.

At night he would tell the children stories. Edgar says, "He would have just gotten us to bed and we would ask, 'Grandfather, tell us a story,' and then he would either make one up or tell us a true one."

Louis was a hero for the boy, for it seemed as if he could make anything with his knife. They would spend hours together in the "whittling" shed, where Louis taught Edgar how to carve a wheel or an axle.

"He was a good shot, too," Edgar recalls, "so we would ask, 'How did you learn to shoot so well?' Grandfather said, 'You have to be able to shoot through fire, is what the sergeant told us.' Then he said, 'When I shot, I always closed my eyes, and I always hit the target anyway.' . . . Those are the nicest memories that I have; I don't have those sorts of memories about my father's father," says Edgar.

Edgar has less vivid recollections of his mother's mother, Thekla—"Omama," as he and Irmgard used to call her. He was five when she died. Edgar does remember going on walks with her to the Gommel—the "Yummel," as they used to call it—Edgar, Irmgard, and the cats making a procession behind the old lady. When she died, Edgar recalls being cuffed on the ears by his father for laughing at the sight of the corpse laid out. He had never seen anything so strange as a dead person before.

Irmgard's first memories are of Omama's death. Thekla died in the Sophienhaus hospital in Weimar but was then brought to Göttern to be buried. Irmgard was not quite four at the time. She has a clear recollection of the coffin being set up where the horse stall used to be and of Frau Schulitz, "the old lady," coming from the village to dress the body. She recalls standing at the window and watching the wagon with the coffin on it, drawn by one horse, pass slowly out of the courtyard on the way to the cemetery. Irmgard remembers crying. She says, "My mother used to say that her [Omama's] death just didn't affect my brother the same way that it affected me. I was more sensitive. My brother doesn't remember very much, and he was a year older!"

It seemed to Irmgard that there was no end to Grandfather Louis's stories and tricks. She says, "We wouldn't go to bed until he showed us a card trick or told us a story. But he never showed us how the tricks worked." Louis had taught his son Walter, however, and the two of them would delight the children for hours. One would put a cigarette in his ear, and the other, sitting across the table, would pull it out from his. Grandfather Louis had a coal stove in his room on which he baked apples for his grandchildren. Irmgard says, "He used to sit there evenings and read the newspaper. It was always warm in there. When you came in, Louis used to look up over his glasses and see who it was. Our other grandfather, Hugo, wasn't as interested in us."

School and the Remarkable Herr Venus

In 1928, when he was six, Edgar began attending the local one-room school located in the Wolfsgrube. In those days, the school year started right after Easter, when parents would be preoccupied with the beginning of the growing season. There were only twenty-eight schoolchildren in the village then, so none of the classes were large, even when two grades doubled up, as was the case with the third and fourth, the fifth and sixth, and the seventh and eighth (the first and second grades were taught separately). Pupils sat in rows of benches attached to a long desk, with a shelf below for storing each child's things. The school day went from 8:00 in the morning till 12:00. At noon, everyone went home for lunch. Three days a week, there were afternoon classes—shop for boys and sewing or other needlework for girls.

For Edgar's first two years in school, Herr Udendörfer was his teacher, an older man who retired in 1929. Edgar says of Udendörfer, "After all those years of teaching, his nerves weren't so good anymore. Teachers were still allowed to hit children back then, and whenever Udendörfer had his problems, if there was only one who didn't obey, he would hit the whole front row of students one after another. They always say, you know, that those who get a lot of beatings grow." Edgar chuckles and says, "I guess that is the reason I am pretty tall." After the war, Edgar paid a visit to his old teacher, who was living in Bad Salzungen. Herr Udendörfer was delighted to see his former pupil again.

Upon the departure of Herr Udendörfer in 1929, the Göttern school underwent a drastic change. Herr Walter Venus came to be the new teacher. What stands out now in Edgar's and Irmgard's minds when they recall their school days is that all eight classes were taught by one teacher. As they put it, "He had to know everything," thinking of Herr Venus and his finesse with the violin.

Born in Weimar, Herr Venus took his training in teacher seminars and attended the renowned music school, becoming adept at the piano, organ, and violin. As the Göttern church organist, he required the children to sing at every Sunday service and at baptisms, weddings, and funerals. For burials, their attendance was also deemed necessary at the gravesite. Irmgard admits to being profoundly moved by the committal of the deceased to their final resting places; she cried as profusely as the mourners themselves. Herr Venus also directed the adult mixed choir, of which Elly was a member. Also an effective dramaturge, Herr Venus organized a theater group for children as well as an adult drama circle, which offered productions during the year that came to be as popular with people from neighboring villages as with local audiences. It was as if he were in continuous motion in the village.

A man of many talents, Herr Venus was also a talented painter. Both Edgar and Irmgard remember staring with fascination at the two large coats of arms painted by Herr Venus for the classroom, one of Göttern (Linden trees) and the other of Thuringia (a lion rampant in a field of stars).

When not instructing his charges in the classroom, Herr Venus was exciting them with treasures to be found outdoors. Göttern, only kilometers from the site of the prehistoric Ehringsdorf man (Ehringsdorf is a village in the outskirts of Weimar), was rich in paleolithic artifacts. Today, however, neither Edgar nor Irmgard can find their beloved collection of tools uncovered in a field under Herr Venus's tutelage. They attribute the loss to the confusion created by the comings and goings of so many people in their house, refugees and others, during and just after the war.

As enthusiastic about botany as he was about archaeology, Herr Venus conducted a "tree school," in which trees and plants were identified and samples of bark and leaves were collected. The children learned to graft trees as well. "He even taught us how to start a fire with stones and with sticks," Edgar recalls.

In a picture in Edgar's possession dating from 1930, Herr Venus stands in the back row with his thirty charges. A handsome youngish man with a bow tie, he appears to be anything but the authoritarian master of his pupils. In the foreground, Edgar kneels jauntily, his black hair cut straight across his forehead, while Irmgard sits discreetly to the far right.

Herr Venus joined the Nazi Party on April 1, 1933. One month later, May 1, he became *Schulungsleiter* (political instruction leader) in Göttern. By the first of October, he was the local leader of the National Socialist Teachers' League (NSL *Ortsgruppenleiter*), a member of the National Socialist People's Welfare Union (NSV *Ortsgruppenleiter*), and a Nazi Party cell leader (NSDAP *Zellenleiter*). Having joined the National Socialist German Workers' Party (NSDAP) ten weeks after it came to power on January 30, 1933, Herr Venus had climbed quickly to local prominence in the party. "He was a fanatic," Edgar says.

"I will never forget him," Irmgard says. "I don't think there was a thing he couldn't do. I was one of his best students, and he came to me and taught me how to write letters. After that, I handled his correspondence for him, while the others were doing their lessons. I had to write his political letters, with carbon paper, by hand, to the party or to [Fritz] Sauckel or whatever all their names were.[2] I learned a lot that way."

Herr Venus was not only the leading Nazi Party personage in Göttern; following his appointment as the keeper of the village council's minutes in 1932, he was the person to be reckoned with in all local political decisions.

Showing the Bulls

At home, Edgar assumed more responsibility with every passing year. Most of his chores had to do with caring for the horses, bulls, and cows. Edwin's decision to purchase a bull for cattle breeding was showing results at the time when Edgar was coming into his adolescence. Edgar learned from his father how to approach, feed, clean, and lead a bull. When he turned thirteen, Edgar was sufficiently skilled to accompany his father to the breeder's auction held in Riesnerstrasse in Weimar.

For Edgar, auctioning the bulls with his father was one of the high points of his life. It meant that a man respected by farmers and breeders throughout the region trusted him with a serious responsibility. The bulls to be auctioned were picked up by Müller, a trucking firm that went from village to village, loading animals. Edgar would ride with them into Weimar and then get the stalls in order. At night, Edgar slept there on straw, to be right at hand should he be needed, while Edwin stayed at his father-in-law's house on Brehmestrasse, near the station.

When auctioning day came, Edgar would use the old toothbrush his mother had given him before he left. She said, "Now son, take this brush and clean up those bulls well." He meticulously went over them, time and again, saving his mother's toothbrush for the detailed work.

Initially Edwin led the bulls around the ring after Edgar cleaned them. Later he allowed his son to do it. A proud day for both father and son came when Edgar, then fifteen, led their prize bull, Kronprinz, tipping the scales at 23 *Zentner*, during the ten-day breeder's show in 1937. After the business of buying and selling was finished, Edgar would take the bulls outside and run them. Later he visited the *Gaststube* inside the auctioning hall and had the "best schnitzel" in town. Edgar says, "You know, we were young then, and we had to eat a lot."

Göttern's First Encounter
with Anthropology

The relevance of good breeding practice, the preservation of breeding lines, the attainment of superior stock, the bread-and-butter of Edwin's life as a successful farmer—lessons he transmitted to his growing son—were brought home to the Schorcht family in 1933–1934 from an unexpected source. Dr. Kurt Bürger came to Göttern that year under the auspices of the Social Anthropological Seminar/Institute at Jena (*Sozialanthropologischen Seminars zu Jena*) and the German Research Institute of Berlin (*Deutschen Forschunggemeninschaft*) to conduct an "anthropological" study of the adult population of Göttern. In the Germany of 1933, anthropology had to do primarily with the investigation of physical characteristics of human populations—that is, with alleged "racial" traits of these populations and their differences—toward the goal of determining degrees of racial "purity."[3] The relevance of such studies was evident to those familiar with Adolf Hitler's polemics on race.[4] The centrality of the "peasantry" to this theme had already been put forth by Richard Walther Darré in his policy of "blood and soil," in which the fate of Germany was seen as dependent on the racial purity of rural folk.[5]

Göttern was chosen by Dr. Bürger as the site of his racial study owing to, as he put it, "its old established peasant population and distance from town and traffic."[6] In the preface, Dr. Bürger thanks "especially the citizens of Göttern for the lively interest they took in my research." To facilitate the project, the teacher and NSDAP activist Herr Venus allowed Dr. Bürger to begin his investigation with the schoolchildren. Then he put himself at Dr.

Bürger's disposal, taking him from farm to farm to introduce him to the rest of the inhabitants. Bürger extends special thanks to Herr Venus in his 1939 book.

Omitting servants and "other strangers," Bürger's results were based on information from seventy-four men and eighty-two women. The data on the children of Göttern were reluctantly excluded to make the Göttern sample comparable with similar studies of adult populations in Thuringia and other parts of Germany.

One of the seventy-four male subjects was Edwin Schorcht, dressed for the occasion, his pictures appearing in plate VII. His apparently "mesocephalic" head rises above a crisp white wing collar, encircled with a silk tie punctuated at the knot by the head of a tie pin.

Irmgard recalls Dr. Bürger's sojourn to Göttern, his visit to the school, the photographs, and the measurements. What remains in her memory most clearly is the assessment of Dr. Bürger's own physique as being short, while it was said of Frau Bürger that she was so tall that "she could drink water from the gutter of a house." Although Dr. Bürger's racially focused research involved Edwin's participation, in reality it constituted a brief interlude in the course of the daily round of the family.

Nazi Youth

By the time Edgar had entered the fourth year of school, he became, like his classmates, a member of the *Deutsches Jungvolk* (German Youth), organized by the NSDAP. It was made up of all boys through the eighth grade. Membership involved learning the verses of "The Horst Wessel Song" and reciting the Nazi oath. Members wore a uniform consisting of a brown shirt, black shorts, a red kerchief, and a dagger. Edgar remembers the store on Streubenstrasse in Weimar where they bought his knife. He says, "The most important thing for us was to play *Kriegspiel* [simulated war games] or this or that out on the fields. And that was fun."

Irmgard joined the *Jungmädelbund* (Young Girl's League) when she was ten. Irmgard says, "Do you know, some of that was really fun. We had a white dress and a brown shirt and a kerchief. Most of the time we did lots of sports, or we sang songs and went hiking. We didn't do any political things. Then, after church confirmation, you automatically became a member of the BDM, the *Bund Deutscher Mädel* [League of German Girls]. We had sports festivals in Mellingen that sent representatives to the regional sports festival. For us, it was quite nice. I was also in the Weimar BDM. It was just as much fun, songs and hiking—it wasn't that political. OK, we had to say, 'Heil Hitler,' but so what?"

Confirmation

Edgar was confirmed in 1936 when he was fourteen. In 1937, Irmgard was admitted into the church at age thirteen, a year younger than was usually the case. Irmgard shared her confirmation with her friend Anita Rausch.

Both Edgar and Irmgard attended the confirmation classes held once a week in Magdala, beginning a half year before Palm Sunday. The week preceding, a test was given in the church in Magdala for aspirants from Ottstedt, Maina, Göttern, and Magdala. The minister conducted the test with a certain forbearance so that even the least well instructed could find his or her way into the church.

The actual ceremony was held on Palm Sunday in the church in Göttern. Irmgard wore a new black dress with just a bit of white, sewn by Frau Spannenberg, who used to come to the house. The year before, Edgar had put on his first pair of long pants and a white shirt. Edwin and Hugo wore starched wing collars and dark suits.

Following the service, a reception for godparents, relatives, and friends was held in the house, preparations for which had begun the week before. On Monday, batter for the cakes was prepared, enough for about three-quarters of the households in Göttern. On Tuesday, the cakes were baked. On Wednesday, Irmgard went from house to house carrying a special platter with a napkin over it, delivering the small packages, each containing slices of several kinds of cake. On Thursday, more batter was prepared. On Friday, cakes were baked for family and relatives; also, the meats were prepared. On Saturday, the house was cleaned, tables prepared, and chairs borrowed. On Sunday, the guests arrived in the late morning to be fed before going to church. The honored guests, the godparents—two men and two women—brought money. In Irmgard's case, they were sisters of her father and mother and two uncles. The other guests came with flowers and small presents—cakes of soap and cologne for Irmgard, a necktie or two and a handkerchief for Edgar. The noon meal was served at one, starting with a good soup, then carp, followed by three kinds of meat, vegetables, salad, fruit, and cream pudding, with copious amounts of wine and beer. It was on this occasion that Edgar publicly drank his first beer. Following the service, the guests returned to the house again for coffee and cakes, with chatting and storytelling continuing into the late afternoon. At 7:00, "evening bread" was served: bread, cold meats, and soup, followed at midnight by coffee and more cake. When the guests left, they took with them small packages containing slices of different kinds of cakes. On these occasions, Elly wondered whether the baking would ever end.

The Gift

Neither Edgar nor Irmgard received remuneration for their work at home, nor did it occur to either of them to ask for anything. All the more surprising, then, was Edwin's presentation to his son one day in the fall of 1936 of a shiny new motorcycle. It was his last year of school, and Edgar, age fourteen, was overwhelmed. He never dreamed that anything so fine would be his very own. The fact that it came from his father, an austere man not readily given to showing his feelings, made it that much sweeter. The motorcycle, a powerful DKW 250, cost 540 reichsmarks (RM). Now Edgar could go where he liked, to Weimar or Jena or even to Bucha—a village to the north where more went on than in Göttern. Edgar had become an enthusiastic dancer and sought out occasions to besport himself on the dance floor.

The Coordination of the Athletic Club

With the completion of the eighth grade, Edgar's childhood was over. Strong and agile, he undertook the work of a grown man on the farm. Membership in the *Deutsches Jungvolk* was followed by obligatory participation in the *Hitlerjugend* (Hitler Youth). "I didn't want to join, but I had to," Edgar says, adding that hardly anyone in the village actually participated. Everyone worked in the fields and had neither the desire nor the time for such things. He did make an effort, though, to participate in the village gymnastics club, which had existed long before the *Hitlerjugend*.

Edgar has fond memories of the gymnastics club, which was set up in a room in the inn where the equipment was kept. The horizontal bar, the parallel bars, and the horse were fastened to the floor. Edwin helped the club acquire them. There Edgar learned how to vault and to walk on his hands on the parallel bars under the supervision of the trainer. Edgar says, "The blacksmith was the trainer. He weighed more than 2 *Zentner*. When he made the 'big circle' on the horizontal bar, the bars would bend."

Edgar later found that his training in gymnastics helped him win recognition in the military. When asked who among the recruits knew how to do gymnastics, Edgar was one of the eight out of nine hundred who identified themselves.

He recalls that it was not too long after the Nazis had come to power that the club was disrupted by the efforts of local party authorities to "co-ordinate" it (*Gleichschaltung*), for the purposes of subsuming it under the program of the *Hitlerjugend*. When that happened, the club changed, and Edgar did not attend anymore. Edgar says, "We always used to say, 'As soon as politics enters a group like that, it goes downhill.' The Nazis cared about assembling a gun, how to clean it, and so on—and that's when the sports part became less important." By the time he was seventeen, Edgar had to join a rifle unit, the *Wehrmannschaft* (a National Socialist paramilitary organization).

Irmgard, Home Economics, and the Villa in Jena

When Irmgard finished the eighth grade in 1936, she was only thirteen. She was a good student; she says, "If it had been up to me, I would have gone to the *Gymnasium* [secondary school] in Jena. I could have handled that. I wanted to learn something different. But my parents said that if I needed to live in the dormitories, they couldn't afford to send me there. Dr. Ochs, our doctor, who gave physical exams in the school, said, 'You should really go to the *Oberschule* [high school].' But then our schoolteacher [Herr Venus], who was a Nazi, said, 'The talented children must be kept in the countryside.'[7] If the teacher had told my father to send me away, my father would have done it, regardless of the cost."

Irmgard must have felt grateful that others saw in her talents that she knew herself to have, potentials that if fulfilled would allow her to transcend the confines of a farming life. That her former teacher, whom she admired so much, allowed his ideological zeal to subvert the possibility of her obtaining a broader education can only have been cause for pain.

A struggle was taking form inside Irmgard between family needs and demands and her own self-fulfillment. She says, "We were born into difficult times, born into agriculture, and one was supposed to stay in agriculture. I could do all the things that were expected of me . . . , but I was never drawn to agriculture. I didn't want to go into the fields; I did it, and I was good at it. I had a very deep love for the animals and for the land, but I didn't want to spend my life out there. My father always told me that his teacher told him that he should be a pastor. With his voice, he could have managed! [*Laughs.*] But he was only one boy with three sisters, and he had to develop into the one who would inherit the farm. He turned out to be a good farmer. Still, I used to think to myself, 'If only your father had become a pastor! Then you could have studied what you wanted to.'"

Although the *Gymnasium* in Jena was out of the question for Irmgard, the Jena Technical School for Home Economics (*Fachschule für Hauswirtschaft*) was not. As Irmgard says, "If that was the route I was to take, then I wanted to be the best I could be. Thanks to my very authoritarian father, I wanted to have the very best education I could have in agriculture and possibly manage a big farm and train apprentices. Father's basic principle was always, 'We live in order to work,' a principle I could never accept. I thought it was more correct to say, 'We work in order to live.' From his code came father's highest claim upon himself and us. We should be perfect in everything!"

Through friends, Irmgard found a room in Jena in a villa with a large yard, lived in by a former diplomat family made destitute by the Great Inflation. Edwin paid for her room and board, relieved that she was to attend a school that made sense. Irmgard was one of several lodgers, including the son of the German consul general in Marseilles, the daughter of a plantation owner in Brazil, and the daughter of a Chinese general. It was, as Irmgard points out, "the" address in Jena. Irmgard says, "I would love to have stayed there forever. That was the environment I enjoyed! That was such a nice year; I learned a lot from them. They were educated people. I was at ease there. They took us with them when they went out to play rummy or to go hiking. I would have liked to stay there to attend high school, but they needed me back home."

Agricultural School

In 1938, Edgar enrolled in the agricultural institute in Weimar. Edwin was eager that his son, heir to the Schorcht farmstead, should receive the same kind of professional training in farming that he had obtained at the agricultural school at Marksuhl. The school in Weimar, located on Schwannseestrasse, had only recently opened and consisted of three rooms. One was for women, where they were trained how to cook for and manage a farming household, while the other two were for men, where they were taught about soil conditions, crop cultivation, physics, chemistry, animal breeding and husbandry, and farm machinery. If Edgar had acquired his knowledge of farming from his father, it was the theory behind the practice that the school would offer him and that would stand him in good stead in the future.

During his two winter semesters, 1938–1939 and 1939–1940, Edgar was instructed in the anatomy and physiology of animals and in the theory of genetics, which enabled him to identify favorable strains in both animals and seeds. He would later attribute to his schooling in Weimar his ability to increase, through enhanced selection of wheat strains, the yield in his two largest fields from 10 loads per quarter hectare to 20 and later 30 loads.

Apart from the specialized agricultural knowledge it taught, the school also made use of an additional teacher for political training. Being a target group of the NSDAP, it was thought appropriate to educate the young farmers in *Blut und Boden* (blood and soil) ideology, the meaning of peasant property, the role of the peasant in the new era, and so forth. According to Edgar, the ideology was not taken too seriously by his friends or by him. Edgar remembers that on the last day of school, the principal, who had lost his leg in World War I, sat on the table, hitting it with his cane and bellowing, "You chaps will become the leading farmers of your communities, and some of you may even be mayors, too." Recalls Edgar, "We laughed, of course, when he said that. But I did become a mayor in GDR [German Democratic Republic] times."

Living in the city of Weimar for the two winters from 1938 to 1940 was an educational experience in itself for Edgar—not because he spent much time visiting the museums associated with so many great figures of German literature, art, and music but because he was exposed to a range of diversity and freedom unknown to him in Göttern. He lived in a boardinghouse run by an old woman at 7 Richard-Wagner-Strasse, right across from the Catholic church, where people regularly congregated in great numbers on Sundays, making a marked contrast to the more casual churchgoing habits of the Protestant citizens of Göttern. Edgar and a classmate of his from the agricultural school shared the house with two students from the School of Architecture, an institution famous for its association with the Bauhaus movement in the 1920s. This was the first opportunity Edgar had to become acquainted with young men his own age who were not farmers. Three times a week, they took their meals together in the boardinghouse living room and talked at length after dinner. On the other two evenings, they would enjoy what the girls had prepared in their cooking class, an occasion for joking and showing off. Edgar says, "That was good stuff. I like to think back on those days."

When the weekends came around, Edgar would race home to Göttern on his DKW 250 to regale the household with tales of his experiences in Weimar, saving time to share with his father information about what he had learned in agricultural school and to work on the projects that had been assigned. No mention was made of politics.

Hitler in Weimar

It was during Edgar's first year in school that Hitler made one of his many visits to Weimar. Hitler and the Nazi Party had a long history of involvement with the city. Located approximately halfway between Berlin and Munich, Weimar was a convenient stopover point; Hitler liked to stay at the Hotel Elephant, made famous by Johann Wolfgang von Goethe's association with it.[8] Hitler made more than forty visits to Weimar, sometimes celebrating his birthdays there and taking time to call on Friedrich Nietzsche's sister, Elisabeth Förster-Nietzsche, a devoted Nazi. But the attraction that Weimar held for Hitler was more than personal convenience—Thuringia, of which Weimar was the capital, provided Hitler with increasing electoral support in the later 1920s as well as important party associates: Baldur von Schirach,

Wilhelm Frick, and Fritz Sauckel. In the election for the Thuringian parliament in 1929, the Nazi Party received enough votes to enter for the first time into a *Land* (state) government, with Frick becoming the Thuringian minister of the interior. In the July 1932 election, Thuringia became the first German *Land* to be governed exclusively by the Nazi Party, under the tutelage of Sauckel, the minister president.[9] Even before these electoral successes, Weimar had been chosen as the site of the second Nazi Party congress in July 1926, held in the same German National Theater where the constitution of the Weimar Republic had been created in 1919, the form of government that Hitler was committed to destroying. In April of the same year, Weimar had served as the birthplace of the *Hitlerjugend* movement under the leadership of Schirach, himself a native of the city.[10]

Edgar was witness to the massive demonstration in Weimar on November 7, 1938, attended by two hundred thousand, at which Hitler intimated readiness for war, proclaiming, "A German is the first soldier in the world or he is nothing."[11]

The agricultural school had prepared its own presentation on that day, in hopes that the *führer* would honor them with his presence. When word passed that Hitler was at the Hotel Elephant, the school emptied in a minute, everyone making their way with haste to the *Marktplatz*. When Edgar arrived, the *Platz* was jammed, and soon the crowd began chanting. Edgar joined in with unrestrained enthusiasm. Over and over they bellowed, "*Lieber Führer, wir bitten dich schön, lass dich noch einmal sehn. Lieber Führer komme raus, aus dem Elefanten Haus*" ("Dear Führer, we beg of you, let yourself be seen one more time. Dear Führer, come on out, out of the Elephant House"). Hysteria reigned in the crowd. Edgar recalls, "We were fanatical then. The youth believed everything, and when the war started, we were fanatical then, too."

Irmgard also remembers that day. She came to Weimar with Anna Rausch to see Hitler and partake of the excitement. It was difficult getting to the *Marktplatz* with the press of the crowd so intent on getting a glimpse of the *Führer*. She never did see him. By the time she arrived, it was said that Hitler had already left through the rear entrance of the hotel.

The Youth Exchange Program and the War

Following the completion of his studies in 1940, when he was eighteen, Edgar entered the Youth Exchange Program. He was eager to see what farming was like in a region where soil conditions were different from those that prevailed around Göttern. Edgar was pleased to be sent to a farm in Rade, near Wittenberg on the banks of the Elbe, while a son from that farm was placed in turn with the Schorcht family. Edgar, accustomed to the hills and valleys of Thuringia, had never seen so much flat farmland before. In Wittenberg, the rich and abundant pastures went right down to the banks of the Elbe. But as in Thuringia, the 36-hectare farm did not consist of all good land; it also contained some sandy soil and some hard, chalky soil. The difference, as Edgar found out, was that these farmers knew how to coax even their most recalcitrant land. For some reason it was not as sensitive to pressure and hardening as the Schorchts' soil was. Edgar wondered whether the presence of the Elbe accounted for the difference. In any event, he found himself working even harder there than he had at home.

In the fall of 1940, Edgar's life as a farmer suddenly came to an end, for while he was still in Rade, he received notice that he was to be inducted into the army on October 3. He left for home immediately.

Commentary

It was not an unwelcome surprise when Edgar presented me one day with a copy of Dr. Bürger's study of Göttern. I had not seen or heard of it before. Looking at it, I realized that in the first part, Bürger's interests are of a historical and social nature, coinciding with my own, whereas the second part suggests, rather, affinities between Bürger's professional interest in race and Edwin's concerns, as a breeder, with bloodlines. As Bürger's research had little to do with the narrative of Edgar's and Irmgard's coming of age, I have postponed a fuller discussion of it until now.

This work of some eighty pages, which finally appeared in 1939, consists of two quite dissimilar parts, the first titled "Folklore." Partaking of the tradition of area studies, it is not without merit, containing as it does original and well-researched information on the environment, local history, the structure of medieval manorialism, settlement typology, economy—the very stuff of modern social/cultural anthropology. This first section bears the weight of demonstrating that the thousands of years of Thuringian history have been ones of successive waves of peoples sweeping "our homeland." The second section, a racial consideration of the adult population of Göttern, as Dr. Bürger writes, "forms the basis of this study."[12]

That such an investigation would have relevance for an intended scientific readership was taken for granted. Nowhere does Dr. Bürger attempt to present the rationale for his investigation—neither the assumptions underlying race as a supposed scientific concept nor the methodological significance of anthropometric measurements. Racial studies of local populations were regarded as self-evidentially legitimate, their justification lying in the very fact of their being undertaken.

Dr. Bürger's methodology consisted of collecting two kinds of evidence, anthropometric (measurements of the human body—in this instance, the head—made by a cephalic caliper) and perceptual, based on three photographs of each subject: full face, half profile, and profile. In the realm of racial studies, the hard data of cephalic indices, reflective of alleged essentialistic characteristics, are complemented by the perceptual evidence of racial type. Racial identification was intended to be apparent to all beholders.

The results of the investigation, laboriously calculated in regard to arithmetic indices, average linear deviations, medium error calculations, and variation coefficients, appear in several concluding paragraphs, where comparisons are made between the Göttern data and similar investigations of seven other populations. In regard to cephalic measurements, the population of Göttern is said not to deviate from the norm, except that "the ear height in Göttern is remarkably low. . . . It is surpassed by almost all the other population groups."[13] Dr. Bürger does not inform us about what possible significance a "remarkably low ear height" might have in the context of racial anthropology.

In regard to hair color and skin pigmentation, Göttern shows a high proportion of "mixed" types. It is not difficult, Bürger argues, to account for the presence of a "mixed race" in Göttern and the absence of a "pure-blooded" population, considering the history of population movements in the area. Somewhat defensively he says, "Of course, in Göttern, there can be found peasants who show a distinctive Nordic Element. However there is no prevalence of the 'Nordic Element' in our results."[14]

In conclusion, Bürger suggests that there is an element of "mental constitution" provoked by race among the peasants of Göttern that determines their initial reaction to this study. In his concluding paragraph he writes, "The peasant of Göttern is of surprising toughness and persistence, which can be seen in his work and his way of life. In the wider surroundings, the peasants of Göttern are known for their reserved manner, for their cautious and rational evaluation of all circumstances and their frugal way of life. It was completely in line with this racial conditioning that our anthropological study at first met with strong distrust in the village."[15]

One can only conjecture about what meaning the Bürger study may have had for Edwin, the breeder of animals. The irony of Edwin's role as a breeder of pure-blooded bulls can only be enhanced by the mug shot of Edwin himself as a subject, in plate VII of Bürger's study. In fact, Edwin's and Bürger's goals converged, insofar as both valued an outcome of breeding perceived as optimal. Only the means for obtaining that common end varied. Bürger, operating ex post facto, worked with the results of largely random mating practices, resulting more often than not in "mixed races." Edwin, as a breeder, controlled the mating environment, pairing sets of prime animals, resulting in the "pure-bloods" that Bürger would have liked to have discovered in human form in Göttern. That a farmer and a research scientist would find, in the human and animal inhabitants of Göttern, a potential intersection of interests is explicable within the environment of Nazi racial thought. In terms of the "logic" of that thought, Bürger's "mixed" findings could be corrected only by following Edwin's breeding practices: optimize the choice of parents. As Edgar put it later on, "One can overlook small defects in an animal if you know that its parents are of the best stock." If careful selection served the animal breeder well, similar practices could be invoked by the racially informed state. It was just such a principle that Darré, an agronomist by training who worked with Heinrich Himmler, adapted in his plans for the racial selection and breeding of a new aristocracy, which was to lead the German people in their expansion eastward.[16]

6

National Socialism

The NSDAP and the Peasant

THE YEARS BETWEEN 1924 AND 1929 were favorable ones for German heavy industry but less so for agriculture, which was undercapitalized and still largely attuned to prewar markets. The world financial crisis of 1929 struck German agriculture particularly hard. Prices for food products dropped drastically, and rye exports virtually stopped. The attempt by the Brüning government in 1931 to insulate the German farmer from world price fluctuations through monetary subsidies proved a short-term palliative. The degree of indebtedness and the number of foreclosures rose markedly. Large segments of the German farming population were ready for a radical solution to their problems.[1]

In the elections of September 1930, the National Socialist German Workers' Party (NSDAP) did surprisingly well in the countryside, mainly in areas of Protestant peasant holdings, winning 102 seats in the new Reichstag, second only to the 143 seats of the *Sozialdemokratische Partei Deutschlands* (SPD). Much of the NSDAP's electoral strength was won at the expense of the conservative *Deutschnationale Volkspartei* (DNVP), which numbered in its ranks members of the Stahlhelm, supporters of Paul von Hindenburg and the Junkers.[2] Such was not the case in Göttern, however, where a Thuringian counterpart of the DNVP, the National Christian Farmers Party (*Christlich-nationale Bauern und Landvolkpartei*), won 90 votes out of a total of 122. The NSDAP received seven, while the DNVP picked up only two.[3] We have reason to think that Edwin, a Stahlhelm member, cast his ballot for the regional National Christian Farmers Party.

Even though the Nazis did not have an agrarian program in place, they had entered the election with a policy statement exalting the peasantry, claiming it to be the "life source of the people and the breeding ground for the armed forces."[4] Such "blood and soil" rhetoric was to become the hallmark of NSDAP discourse in the months and years ahead.

The NSDAP reacted swiftly to exploit its electoral advantage among farmers. The architect of the NSDAP agrarian program was Richard Walther Darré, an Argentinean by birth, trained in agronomy and zootechnics at the University of Halle. Darré became known for his papers on livestock selection. Though schooled in science, he was immersed in a pastiche of racism and naturalism. In 1929, Darré had written *Das Bauerntum als Lebensquell der nordischen Rasse* (*Peasantry as the Life Source of the Nordic Race*). In 1930, a second book, titled *Neuadel aus Blut und Boden* (*New Nobility from Blood and Earth*) was published. In that book, Darré envisions a racially regenerated aristocracy of peasants and warriors, created by rigorous selection capable of restoring the "traditional values" of German culture.[5] These ideas were to lead Darré into close contact with Heinrich Himmler and the *Schutzstaffel* (SS). Together they assumed the task of uncovering the alleged selective mechanisms for making the SS, in concert with the peasantry, the pillars of the new outposts in the "eastern regions," to be won from the Slavs.[6] As Karl Bracher points out, the dialogue between Darré and Adolf Hitler (and Himmler) dealt only with the question of "how agriculture can be used in the present battle for the state."[7]

Toward this end, Hitler had commissioned Darré to organize an "agrarian apparatus" (*Agrarpolitischer Apparat*), or AA, within the NSDAP in July 1930, two months before the election, to mobilize the countryside. This was done by infiltrating and taking over existing traditional representative organizations, such as the *Reichslandbund* (German Rural League), and in their place creating a vertical structure, with local offices at the county and the village levels.[8] The AA was officially dedicated in Hitler's presence on February 8, 1931, in Weimar, two years before the accession of the NSDAP to power. From 1933 on, the AA ensured the execution of the regime's directives by the local population. In the county of Weimar, as elsewhere, the AA hierarchy consisted of a county peasant leader, a county manager, and three county representatives, while in Göttern, the AA was represented by a local peasant leader.

Darré's organizational efforts on behalf of the NSDAP paid off handsomely in the Reichstag elections of July 1932, making the NSDAP the strongest peasant party in Germany.[9] In Göttern, the NSDAP won fifty-three votes to sixty-six for the Thuringian Rural Association (*Thüringer Landbund*), demonstrating how enduring conservative voting behavior could be in a small village. Only one vote was cast for the left, the SPD.

The proportion of votes for the regional party, the Thuringian Rural Association, was even higher in Göttern in the elections of November 1932; it won sixty-four votes out of a total of ninety-six votes cast, as compared to twenty-six for the NSDAP.[10] In the presidential elections of April 1932, although Hindenburg decisively defeated Hitler, Hitler's national vote of 13,418,000 to Hindenburg's 19,359,000 was hailed as a moral victory, for it showed that the party's strength had nearly doubled since 1930.[11] In Göttern, however, the traditional right did proportionally better in the second round of the presidential election (59.6 percent) than it had at the national level (53.2 percent). Hindenburg received fifty-six of the ninety-four valid ballots cast, to Hitler's thirty-seven. One vote was cast for Ernest Thälmann, the Communist Party candidate.[12] As a Stahlhelm member, Edwin undoubtedly cast his vote for Hindenburg as opposed to Hitler. The Reichstag election of March 5, 1933, was the last opportunity Edwin would have to vote his choice, a right granted to him by the vilified Weimar Republic.[13] It was undoubtedly Hitler's decision, now that he was in power, to allow this election to demonstrate overwhelming support for him. Out of a total of 146 votes, the Fighting Front Black-White-Red (*Kampfront Schwarz-Weiss-Rot*)—the conservative successor to the Thuringian Rural Association—continued in the majority with seventy-five votes, even though its lead over the NSDAP (sixty-eight) had narrowed considerably.[14] In the Reichstag elections of November 12, 1933, following the suppression of all other political parties in the summer of that year, the NSDAP received all of the 152 votes cast in Göttern.[15] By then, no opposition to the hegemony of the Nazi Party was possible.

In May 1933, after the NSDAP came to power, Darré was made Reich peasant leader, and in June, minister of Food and Agriculture upon the resignation of Alfred Hugenberg, signaling the end of Nationalist control of agrarian organizations.[16] In the fall of that year, Darré promulgated a number of decrees, which were to serve as the centerpieces of NSDAP agricultural policy.

The Reich Food Estate (RNS)

His first decree, known as the Reich Food Estate (*Reichsnährstand*, or RNS), was issued on September 13, 1933.[17] It was designed to unite in a single immense corporation all producers, processors, and distributors of agricultural products and foodstuffs. Based on Darré's absolute control leadership principle (*Führerprinzip*) and obligatory membership, the concept of the RNS drew on corporatist ideas, operative in Fascist Italy, for the solution of problems of production and distribution rather than reliance on the market forces of supply and demand.

Starting in 1936, when the RNS became fully operational, farmers, depending on the size of their farms and the quality of the soil, were assigned annual delivery quotas, particularly for milk and grain. The RNS was able to control the production and distribution of all food products and fix their prices as well.[18] The concept of a "fair price" was an ideal close to Edwin Schorcht's heart and constituted one aspect of the RNS that he undoubtedly favored.

The Reich Entailed Farm Law

The second major decree issued by Darré, one designed to implement the mystique of "blood and soil" (*Blut und Boden*) as well as fulfill the dream of "re-agrarianizing" Germany, was the Reich Entailed Farm Law (*Reichserbhofgesetz*) of September 29, 1933.[19] The purpose of the law, in keeping with the party's racial goals, was to ensure the perpetuity of the "bloodline" on the soil by controlling inheritance while at the same time withdrawing the property from the open market by prohibiting its sale or being mortgaged.[20] To this end, the preamble states, "The Reich government desires the maintenance of the peasantry as the source of new blood by safeguarding old inheritance customs which prevent over-indebtedness and parcellization of the family holdings in the hands of free yeomen."[21] The inheritance law was the cornerstone of the vision of a new nobility (*Neuadel*) to be drawn from the German peasantry.

Specifically, the law restricted the status of "inherited freehold" (*Erbhof*) to a farm large enough to provide a family living, at least 7.5 hectares and not larger than 125 hectares.[22] Only a person who possessed such a farm and worked it himself could be called a peasant. The distinction was made between peasant (*Bauer*) and farmer (*Landwirt*), the former seen as representing the true *Volk*, the "new nobility." The latter, J. E. Farquharson suggests, belonged to the French Revolution and self-seeking liberalism.[23] The farmstead must be bequeathed undivided to one son, according to primogeniture inheritance rights. The *Erbhof* was free of inheritance and real-estate acquisition taxes. In return for protection by the state from "Jewish-implanted" market forces, the *Bauer* was committed to placing the needs of the *Volk* above his own, preventing him from alienating or speculating on his inherited freehold for his own purposes. Placing ideology before economic need, Darré insisted, "A farmer must always consider his activity as a duty towards his family and his people, and never simply as an economic venture from which he can earn money."[24] It was, as Farquharson points out, as if the entailed peasant had been placed under a new form of neofeudal obligation, the medieval lord having been replaced by the *Volk*.[25]

The Schorchts and the NSDAP Agrarian Policies

Despite the discord that the Darré program provoked among some of the German farming population, the Schorchts perceived the 1930s as good years. Edgar says, "Apart from the political side of things, the economy of the Hitler years worked in our favor. The silos that we were able to install for subsidized costs, two for dry fodder and one for steamed potatoes, are proof of that." Also between 1936 and 1939, Edwin added a number of important pieces of farm machinery at favorable prices: a McCormick hay tedder, a double plowshare, a harrow, an electric turnip slicer, an electric saw, and a liquid-manure pump.

The energy and thought that Edwin invested in the scrupulous care of his fields and in his livestock ensured that he was able to readily meet the production quotas imposed on him by the RNS. Like all livestock farmers, Edwin was beset by high feed costs, but the carefully planned breeding program he started in 1928 came to fruition in the late 1930s, just when Germany was confronting a severe shortage of animal fats.[26]

The Schorcht farm became an *Erbhof* in 1935, a fact that had little immediate consequence considering that entailment—limiting inheritance to the oldest male child—had been practiced by the Schorchts in Göttern since settling there. By using de facto entailment as long as they had, the Schorcht farm had remained intact through five generations. If Edwin did have reservations about the power of the decree to limit his authority, he undoubtedly kept it to himself. Edwin was a patriot and most likely considered it his duty to obey the government.

What did weigh on Edwin's mind, though, was the effect restricted inheritance had for children not destined to inherit the farm. Since that September day in 1922 when Edwin had driven to the Blankenhain courthouse with his father, his three sisters, and their husbands to legally become sole owner of the farm, he had had to live with the consequences of that official act. It had forever tainted his relationships with his sisters, for either they or (more likely, from what he could tell) his brothers-in-law were never satisfied with the agreed-on terms of their compensation. His initial attempt to meet the demands of the settlement in inflated reichsmarks (RM) had been discredited. As it was, Edwin believed that he would never be able to make it up to his daughter when the time came to settle the farm on Edgar.

The Family Tree

In applying for the status of an entailed farm for the family property, Edwin was obligated to account for the racial purity of his ancestry to two ascending generations. Edgar, his son, was required to perform the same task on two different occasions in 1939: first to meet the demands of the RNS, and then in preparation for writing his booklet *Der Bauer und sein Hof* (*The Peasant and His Holding*), undertaken as a requirement for graduation from the agricultural school in Weimar.

The first was prepared on a special "Family Tree" (*Sippschaftstafel*) table, bearing the blood-and-soil insignia of the RNS on the cover. Beneath its title, the form bears an admonition written by Darré: "Keep your blood pure, as the holy legacy which the God of this earth has bequeathed to you." In addition to giving the names and residences of his parents, grandparents, aunts and uncles, and first cousins, Edgar was required to name their occupations where relevant, dates of birth, dates of death, heights, skin colors, and hair forms and colors. In some instances, spaces are left blank where Edgar lacked information, but that was obviously not of great import. What is clear is that in Edgar's immediate lineage, there is no trace of "impure blood": no Jews, no persons of Slavic origin, no obstacle to prevent him from being an entailer.

For the second table, which appears just inside the cover of his booklet, Edgar was required to provide information on his parentage for three ascending generations, although no data were prescribed for collateral relatives, aunts, uncles, and cousins. Again, the name Edgar Paul Hugo Schorcht, as entailing son, appears with no space left for designating siblings or, in this instance, a wife. The Reich's perception of women was unequivocal: their function was to bear children for the *Volk*. The entailed son's responsibility was to ensure the transmission of the inherited "soil" intact, as a crucible for the "blood."

Additional proof of genealogical purity was required of Uncle Walter, Edgar's mother's brother, a minor official in the NSDAP. In the Schorcht family circle, Walter Michel is still known as the "Nazi" uncle. As a young man brought up in Göttern, Walter might well have stepped into his father's shoes as keeper of the family inn had Louis not sold it. Without a clear idea in his mind for the future, Walter left Göttern for Weimar, fertile soil for a young man who yearned to make something of himself in the NSDAP in the early 1930s, away from the strictures of small-town life. Soon becoming an official in Weimar's office of party finances, he associated with party luminaries, including Fritz Sauckel, entertaining them in his father's house on Brehmstrasse in Weimar.

Uncle Walter returned to Göttern from time to time for family occasions, but he did so at some cost to family harmony. Irmgard says, "My father and my uncle were always politically at odds. Whenever there was a family reunion, things became difficult. When they got started, my mother would say, 'Oh God, they're going on again about politics.' They were like two fighting cocks. My father stood here [*pointing*] and my uncle there, and they went at it. My father was a Stahlhelm. He always said, 'What you people have started just cannot end well.' Uncle Walter went on talking and talking about the *Volk*."

During the war, Walter Michel was given a promotion and sent to Zell in the Rhine Province. According to Edgar, most of the family in Göttern was not sorry to see him go. He recounts, "We used to say, 'Thank God he was shipped to Zell an der Mosel,' but even there he was still a big Nazi."

The genealogical document used as evidence of Willy Max Walter Michel's Aryan background shows a red check mark placed next to each entry for the nineteen ascendant relatives by the official into whose trust the document was delivered on July 1, 1943. When Walter Michel was arrested by the Americans two years later, the form that he had so painstakingly worked on to demonstrate his genealogical worth served as one more incriminating piece of evidence of his Nazi credentials.

Edwin's Involvement with the NSDAP

Despite Edwin's voiced antipathy to the Nazi Party, as recalled by his children, we learn from the records of the Göttern Village Council meetings, kept by Walter Venus from 1932 to 1945, that Edwin was a council member in 1932 and remained one until 1934.[27] This comes as a bit of a surprise in regard to a man we have known primarily as a farmer. Neither Edgar nor Irmgard said anything to imply that their father had any particular interest in politics. Indeed, his apparent formality suggests an aversion to the give-and-take associated with it. Yet in addition to being a member of the council in 1932, he reappeared in public life in 1944, when he was nominated by Venus to be first assistant mayor.[28] On April 18 of that year, Edwin, in conjunction with the new mayor, took the oath of office in the presence of Venus, swearing loyalty to "the leader of the German Reich and the German people, Adolf Hitler."[29]

That their father was an NSDAP member was, however, made mention of by his children. According to Edgar and Irmgard, he became a party member "toward the end of the war." His signing up is regarded as a sacrificial act, committed for their sake rather than as evidence of political or ideological conviction. It was, they insist, the constant badgering of Venus that won him over. Edgar says, "Venus would exclaim, 'If you don't join up, your children will be disadvantaged when the war is over.'" Irmgard adds, "Finally he gave in, saying, 'Do what you want; just leave me alone.' So he paid his membership dues."

A copy of Edwin Schorcht's NSDAP membership card from the Bundesarchiv in Berlin shows that he became a party member on April 1,

1942. Whether that date can be considered "early" or "late" is a matter of perspective. What is clear is the desire of Edwin's children to protect the image of their father as a principled man.

For Edwin, the outcome of the war may have been fateful in any case. As Irmgard points out, "My father said, 'Either way will be terrible, for even if we win, we won't see our children again; they will be lost to us in the East somewhere.'"

The POWs

It is not possible to know exactly when the Polish prisoners of war (POWs) came to work for the Schorchts. The fact that they are listed in Edgar's *Der Bauer und sein Hof*, distributed in 1940, suggests that they had been rounded up by the Germans in the course of the invasion of Poland. They were, according to Edgar, sent by train to Weimar and, once there, distributed by the *Bauernverband* (peasant association) to those who needed farm labor. The Schorchts were obliged to provide them with room and board in addition to a small cash wage.[30]

The Schorchts had a Bulgarian working with them in 1938, who helped build the new barn that Edwin had put up on the piece of land across the way. Edgar says, "He didn't know how to work, though. Perhaps that was because he came from a warm climate. He told us that his brother was a sheepherder. In any event, he stayed until 1940."

Polish POWs lived and worked for the Schorchts for at least four years, including Sigmund Paprizki, who came from Lodz knowing not a word of German. Paprizki learned to speak German quite well while living in Göttern. He was outgoing, quite unlike Bruno Jenziwski, who was reserved and distant. Edgar says, "Bruno was not friendly, certainly not a Germanophile. Well, after all, you can't really ask for that. As a prisoner—you can't demand that he be friendly with the people whose prisoner he is. But Sigmund became a good friend. Then there was Sophia Sarembo; she knew our folk songs better than we did."

The Poles lived in the small rooms across the courtyard and ate their meals with the family, an arrangement that was strictly forbidden by the government. According to Edgar, his father was adamant about their presence at meals: "'They eat here with us because they are human beings,' he would say. There were controls, people from Göttern who became 'little Hitlers,' running around to check to see if the Germans ate at the same table as the prisoners, but my father always insisted on that and defended himself for doing so."

Irmgard remembers that it was her father who told the other farmers that he was going to see that the workers got home to be with their families

for Christmas. She says, "Everyone said, 'They'll never come back. Don't let them go.' But my father wanted them to be with their families. He succeeded in organizing it for Göttern. Sigmund got to visit a child of his whom he had never seen. We sent toys with them, also with Bruno, our other worker. They left and came back, on time as prearranged."

Although the Poles were of prisoner status, they were free to come and go within the village limits. Between eight and ten were lodged at the old Dobermann estate, leased by the Wagner family. Stevka, a young Polish woman who was entrusted with getting the mail, lived there as well. Edgar says, "Every day she hitched up a cart and horse and drove to the train depot at Grossschwabhausen and picked up the mail. She was given that trust, and nothing ever happened." In the evenings, most of the thirty POWs in Göttern would congregate at the Wagners' place, under the supervision of guards.

Of the POWs, Edgar says, "The prisoners in Göttern weren't really prisoners, because they could walk around as they wished and only had to wait until the war was over to return to their homelands. They were free, but we had to work, and they worked with us, one better, one worse—it's always like that." These words seem to sum up Edgar's understanding, if not rationalization, of a piece of family history about which he has some compunctions. What can be salvaged from it for him are the friendship with Sigmund and the pride taken in his father's insistence that the POWs eat with the family.

Irmgard and the War Years

For much of the war, Irmgard was away from home, despite the fact that her presence was needed on the farm. In 1941–1942, she enrolled in a home economics program in Ronneburg, just east of Gera. She was eager to get started on a career. After the first year, she hoped to stay for another so that she could qualify to be a teacher of agricultural apprentices. Irmgard says, "My teachers said that they were so happy to see my name on the registry list. Then my father wrote that he had been struck by a horse, Edgar was fighting in Russia, and I should come home because he couldn't drive the tractor anymore. His knee was injured, and he didn't have anyone else. I had to come home." Irmgard believed, she recalls, that duty to farm and family should come first, regardless of the cost to her own ambitions.

Living and working at home then was not the way it had been before Edgar had gone to war. Both of her grandfathers were now dead. (Irmgard had received word on January 21, 1942, that her grandfather Louis had died, having fallen down the stairs and had a stroke. He was eighty-one.) Irmgard remembers having a frightening dream about them. She says, "My

mother always told me to keep my dreams to myself, because I often had dreams about things that later happened. In my dream, I wanted to go into the kitchen, and there was a deep hole, like a grave. One grandfather was in there, and the other one was on the edge. The one in the hole was beckoning the other to come down and join him. I don't remember how the dream ended, but I was afraid to go into the kitchen—I was so scared!

"It was so cold that winter. My grandfather Hugo didn't have a stove, so they put him down in the little *Stube* so that he would be warm. He wasn't sick; it was just so that he could stay warm. My mother came in to stoke the coal and asked him how he was. He said, 'I can't go to Louis's funeral today; it is too cold.' She told him, 'Of course, you don't have to go. Just stay right there in your bed.' Mother left to milk and came back to find that Hugo had died. Louis wasn't buried yet, and Hugo had died! My dream, do you see? The one took the other away with him!" Irmgard remembers that it took a long time to dig the two winter graves, because the ground was so hard.

Even though agriculture was not her chosen occupation—she remembers having nightmares about working in the fields forever—Irmgard was determined to advance herself as far as she could. To complete her program, she needed to apprentice at another farm. This was in 1944. Irmgard says, "I found one [farm] near Magdeburg from a trade journal advertisement. I had sent them my grades—they were excellent. On this farm, the husband was away at the front, and just the wife and only one of the helpers were left behind. It was a big farm. I would have liked going there. But then my mother said, 'What, my boy is in battle, and we don't know if he is coming back. Now you want to go far away from me? And what if the bombs come?' So once again, I gave up something good for me for my parents' sake and went to a little farm near Apolda owned by some relatives. It was smaller than ours. I didn't help in the fields; we worked in the garden. The woman didn't work as hard as our mother. I might just as well have stayed at home. There wasn't anything new for me to learn there."

When Irmgard finished apprenticing at the farm in 1945, she returned home to Göttern. Her mother was hard-pressed getting the housework done, to say nothing of the milking. It helped when the Raiffeisen Cooperative in Magdala bought a large washing machine. Members were able to reserve a time to do their wash there. Irmgard says, "We prewashed everything at home, and then I usually drove the horse and cart of laundry to Magdala. We washed and hung everything out to dry, and then that same day we pressed it. That was a lot of work.

"There was a group of women known as the *Spinnstube*. My mother knew lots of women in Magdala—the pharmacist's wife, the doctor's wife, the brewer's wife. My mother always had interesting people around her. Once a week, they all got together at one of their houses and mended clothes.

They sewed, drank coffee, and caught up on the news.[31] These women used to laugh at me because I would come by bike, then I was on a tractor, then next time with a horse and cart, then with my motorcycle!"

Wartime and Food

It is generally agreed that the German population was relatively well fed during World War II, at least until 1944.[32] Part of the credit for this goes to the RNS and its system of obligatory production quotas. Acute shortages and severe deprivation did not occur until after the war, when the serious dislocations ensued. The least-affected population, even in 1945, was the peasants. Irmgard says, "We didn't notice any real food problems in the village during the war. My father always used to say, 'When the farmers starve, then the others have long since died of starvation.' Everything was rationed, sugar and meat and butter, and we could slaughter only with permission. There were some who could not slaughter at all, because they had not filled their quotas. But we had potatoes, vegetables, poultry, and rabbit. At one point, we had fifty rabbits. We used the fur for shoes and for clothes. The city people would trade us sugar or yeast for meat or vegetables. One woman came all the way from Leipzig. She brought yeast and sugar, stayed over at our place, and then left with provisions from us the next day."

News of Edgar

What concerned Edwin, Elly, and Irmgard as the war years wore on was not the prospect of want but the safety and welfare of Edgar fighting on the Eastern Front. They wrote to him frequently—all three prided themselves on writing a good letter. Occasionally, they would receive a reply via field mail. Not allowed to say where he was, Edgar had made up a special code for his father. He put a dot above certain letters so that Edwin could spell out the name of the city where he was and follow his movements on a map of the Soviet Union. The family would send so-called pound packages (500 grams), usually containing cookies and an occasional handkerchief.

Irmgard says, "My mother and I really cared for him. When he left for the front, we said our good-byes. I felt a bolt go right through me, and I knew. . . . I said to my mother, 'He'll be back.' I felt it even when we didn't hear from him for such a long while."

The last time Edgar had been seen was in a base hospital in Freudenstadt, near Freiburg im Breisgau in the Black Forest, where he was recuperating from a wound suffered when attempting to detonate a mine. Despite the difficulty and danger of long-distance travel, it was decided that Irmgard would visit him for his birthday on January 25, 1944. Irmgard was excited at the prospect of her first long trip. She took a cake with her.

Commentary

What a priori questions I had about how an ordinary German might undergo the war years and come to terms with the war itself were readily answered by the Schorchts' memories of that time. They were well off as compared to most city dwellers and, as farmers, could trade what they had in abundance for any items they lacked. The manpower shortage was alleviated with the availability of Polish POWs. Sigmund and Bruno proved to be good workers, with Sigmund's conviviality more than making up for Bruno's taciturnity. The family did not experience the terrors of an air raid; rural villages were relatively safe places in which to live out the war years. They were undoubtedly together in their concern for Edgar's welfare and apparently without reservation as to the rightness of the German Army's cause. This was Edwin's and Elly's second world war. What they had probably learned from the first—an "unjustified" defeat and the imposition of the "draconian" Versailles Peace Treaty—may well have sanctioned in their eyes the second, establishing a reasonable connection between them.

More of a challenge was Walter Venus, if for no other reason than the impact he had on members of the family at different times. The more I came to know about him, the more provocative I found him to be. Venus was a man of seeming contradictions. Either because of or despite his talents as an inspiring teacher to young and old alike, he was instrumental in furthering the acceptance of National Socialism in Göttern. Joining the party in April 1933, he rose rapidly to prominence in local circles. He employed every opportunity to further party interests and activities compatible with its ideology—for example, providing Dr. Kurt Bürger with permission to start his racial anthropological research in the Göttern school, using the students as his first subjects. Venus was also assiduous in recruiting village residents for the party. He was, according to both Edgar and Irmgard, a "fanatic."

It may well have been at this time that Venus, in company with others, began to see National Socialism not only as a solution to the professional crisis facing teachers in the early 1930s but as a path to a political career as well, with access to the authority and power afforded to young men by the party.[33] Walter Michel, Elly's brother, had already come to a similar conclusion, having become a party member in 1931 to supplement his career as a revenue officer with party activity.

That Venus experienced no apparent conflict between the aims of teaching and the promulgation of the exclusionary aims and elitist ideology of the NSDAP finds confirmation in the realization that medical doctors, on Hippocratic grounds, equally if not more committed to humane goals, joined the NSDAP twice as frequently as did teachers.[34] It may be that Venus held a reciprocal view of the relationship between teaching and party mobilization, believing that his skills in teaching supplemented and lent credibility to his commitment to the party, and that his love of the arts helped adorn the party he loved. Indeed, it may be an error to separate these two domains of Venus's life and place them in opposition to each other, as if their apparent antagonism represented a flaw of character. Teacher and Nazi Party functionary he was, yet Venus was also an organist, a naturalist, a dramaturge, an administrative functionary—a "man for all seasons." Seen from this perspective, Venus's marriage of political and pedagogical competencies would seem to constitute less of an anomaly.[35] Nonetheless, this perception of the exceptional integrated persona, a "Nazi Humanist," cannot be sustained through 1944–1945, when the fires of the Holocaust were lit and maintained as much by teachers and doctors as by nonprofessionals.

7

Edgar at War

A Sapper

EDGAR RETURNED HOME from Rade on September 30, 1940, and reported to Weissenfels (35 kilometers southwest of Leipzig) on October 3. There he found out that he was to be trained as a sapper, someone who lays and disarms mines. Despite his apprehension, there was no doubt in Edgar's mind that that was what he should be doing when the moment came to be inducted. His father and his grandfather Louis had lectured him about the importance of military service. Edgar remembers being taken to Leipzig by his father to be shown where he had trained before being shipped off to France in 1914: "My grandfather and my father said, 'You have to become a soldier; that's where you are really trained to be a man.'" He went home to pack his things.

Edgar says, "I got home around two o'clock in the morning. My mother was looking out the window, waiting for me. She said, 'There you are, my son! You are surely going to have to serve. What do they want you for?' I told her, 'assault sappers' (*Sturmpioniere*). My mother called out, 'Oh no, not that!' She started to cry. I tried to comfort her. I told her that we had it better than the guys in the motorized artillery. We could always run away quick, while they were sitting ducks. It's hard to bomb a lone man. Actually, it was a suicide mission," Edgar says. "As sappers, we were taught the worst stuff you had to do in the war. We went out in front of the infantry to set up barbed-wired fences and lay mines. We told the infantry where we were; otherwise, they would shoot us too."

The training at Weissenfels was intensive. The sappers learned how to construct bridges and how to mine them for destruction. At one point, they built a dam over the Saale River, creating a crossing some 80 meters wide, over which the whole battalion had to build a pontoon bridge. In the winter, they constructed an ice bridge from ice blocks over which water was fused to seal the joints. First foot soldiers were ordered to cross it; later, a truck was driven across. After that exercise, there was little doubt in the minds of Edgar's training group that they were to be sent to the Eastern Front.

The Soviet Front

In April 1942, Edgar's unit was moved east to the Soviet front and into action in the Stalingrad sector. There, just before going into battle, Edgar was struck with appendicitis. He was operated on in a field hospital 80 kilometers west of Stalingrad. Edgar realized what he had missed when the sole survivors of his regiment—a mere twenty-three men, all wounded—made it back to the hospital. Edgar says, "My appendicitis kept me out of Stalingrad. The doctor didn't know where my unit was; I knew, but he didn't."

Edgar's appendix scar had hardly healed when a hernia was discovered in his left groin. He had apparently had it since the beginning of his training, building bridges in Weissenfels. Once more, his "good fortune" spared him. Edgar was first put aboard a freight train to Kharkov near Kiev and then put on a hospital train to Lublin. At the university hospital, they determined his blood to be too thin for surgery. To his great relief, he was shipped ever farther west, finally all the way back to Weissenfels, where he had started. Edgar says, "I got transferred farther and farther away from the front. Those doctors were good people back then, not like the SS. *They* wouldn't have done that for me." There, with his blood now coagulable, a surgeon successfully operated on January 12, 1943, as the battle raged in Stalingrad.

Edgar was given a convalescent leave, and he returned home. The reunion was joyous, and Edgar relished every moment of being taken care of by his mother and sister. Never had his mother's dumplings tasted better. This reprieve, however, was short-lived.

Wounded

Following his recuperation, Edgar was returned to the Russian Front, where he became a casualty for the first time in the fall of 1943. The Russians had taken to enclosing their mines in wood to disguise them and equipping them with a release valve activated by 10 pounds of pressure. It was in the process of attempting to remove a mine encased in heavy mud that Edgar discovered how easily they could be set off. The blast blew him off his feet and onto his back. At first, he didn't know where he was hit. When he raised his arm to wipe the blood from his right eye, he saw the wood splinters protruding from the back of his hand. He was removed to a field hospital.

Within three months, he was deemed combat-ready and reassigned to the same zone, where he was wounded for the second time, now by metal grenade splinters. The damage was more severe; there were wounds in his legs as well as in his hands. This time, his convalescence took longer. At first, he could walk only with the aid of crutches. Edgar says of the field hospital: "When we were better, we went into the marching company. When we made it there, it meant we were ready for war again, and we were shipped back to the front."

The SS

"There we ran into the SS, Hitler's boys. They gave the orders, and we in the Wehrmacht had to jump. We didn't like them. If a guy has his hands up in the air, then you don't shoot him anymore. We would round up the enemy and be on our way to take them to a detention camp, when the SS would come along. They would take our prisoners somewhere and shoot them. Later on, we were in Kiev, and we even saw where they took the Jews. The Jews were the first ones rounded up. There were labor camps where they were worked to death. That wasn't that late, maybe 1942–1943. But you couldn't say anything. We guessed what they would do to them. We knew there were ghettos in Warsaw and a couple in Kharkov and Kiev. But as an ordinary soldier, we couldn't do anything about it."

The Soviet Countryside

"We marched almost everywhere, up to 30 to 40 kilometers a day. We used to say it was like what they did to Napoleon. The Russians suckered Napoleon deep into Russian territory. They retreated farther and farther and left only old people. So it was with us: women and children at the front, and the troops deeper and deeper into Russia. The young people we found were deported back to Germany to work. They worked for us because we [men] had all been called up to battle.[1] There were days when we had to look for young children when we came into a village. We would go from door to door, looking for children. Once we came across a woman, and she was crying and crying, '*Moya, moya, malenka?*' ('My son, my son, where are you?'). He had hidden himself. We told her not to worry; we wouldn't reveal her secret. She was so happy; she gave us lots of berries and something to drink.

"We were in that part of Ukraine through which the Donets River flows toward the Don. I saw there the most fertile soil I have ever seen in my life.

We still had small plots of land that we worked by hand, and they already had huge tracts of land. Their rye was much better than anything we had ever seen. Their soil was so fertile; we could only dream of such soil at home. No stones. And what was really strange for us—we would walk through fields that they had cultivated, and then for hours at a stretch we'd walk through fields that hadn't been tilled at all. They could afford to leave land fallow because they had so much. What I could not understand is why the area around the fields was not better maintained. They planted potatoes, rye. They didn't have silos, like we did. When they harvested, they just piled stuff up on the fields. The stuff on top rotted, but underneath was OK. That was how they kept their feed. Their cows grazed outdoors all summer, and in winter they went into the stalls. But it looked like they weren't very good at storing feed for the cows. I doubt if all of them survived the winter. That's how I saw it as a farmer. We saw them milk—they had so much milk they didn't know what to do with it. They would give us bucketfuls. That took care of the diarrhea we had from their water!

"The people were scared of us, but we never did anything to the civilians. I was always a little cheeky; one day I smelled something fine cooking in someone's house that smelled so good—I went in. I pointed to their food, sniffed, and said, '*Karacha, karacha*' ('Good, good'). They grabbed a plate and served me some, and I said, '*Spaciba, spaciba*' ('Thank you, thank you'). And then I went on my way. Once we saw a woman—that was kind of dangerous, to stay behind as a woman, but we never did anything like that. We had to go out one night to lay mines, and just around then we got a change of clothes, which was rare. So we washed out our clothes, hung them out to dry, and went off to lay mines. When we came back, our wash lay out on the table, ironed and folded. The woman had ironed out our clothes. She had a coal iron; they didn't have any electricity in those villages. We gave her some fish or preserved meat as thanks. I know that when I was operated on in Millerovo, a Russian woman from the village came to wash for us. Her son was off in the battle too. I talked to her a little bit. She had to wash for the enemy."

Edgar Becomes a Sharpshooter

"When we finally got to our sector, we found that every Tuesday, there was shooting practice on the range. One beautiful day, when the orders were given out, they read the names of twenty-four of us who were supposed to step to the side. The rest were being shipped off to the front. We were worried about what they had in store for us. They had taken all the men who were good shots on the range. I had probably learned something about it from Grandfather Louis. They said they planned to train us to be sharpshooters.

And we said, '*Scharfschützen? Das Himmelfahrtskommando machen wir nicht mit*' ('Sharpshooters? No, getting on the fast track for heaven is not for us'). But they didn't ask us!

"So there were the twenty-four of us, a teacher, and an aide. Our teacher was also our company commander, a fine man. That was the best training I ever had. Earlier, he had worked in a factory as a fitter. We were a loud, rowdy crew. Training lasted eight weeks, and after that we went to the training field in Zeithain near Riesa. There we had another training course, where each of us got a rifle that later we would take into the field."

The Last Battle Ground: East Prussia

"We were given a choice of two fronts, Lithuania or East Prussia—that's all that was left! Lithuania already was filling up. Two of us decided, 'Well, let's go to East Prussia.' But when we got there, we were never deployed as sharpshooters, even though we wanted to be, fools that we were. When I think about that now . . . why, why?

"By then, we were old scouts. The Hitler regime was taking really young people who had had four to six weeks of training. They could hardly hold their weapons and were supposed to lay mines, shoot antitank guns. They were either scared or curious. . . . The older fellows were quite careful and peeked around corners; otherwise, you would get it from a sharpshooter. The new ones who came—within a week, half of them were already dead or wounded. That's when it was over for Hitler; we didn't believe in it anymore. By the middle of 1944, the war was already lost. Once an army has retreated, it can no longer take a good stand. That's how it was up in East Prussia. We couldn't hold ground after that; we were a slaughtered herd. The Russians were also beat, but they had so many men from our side—either those who had had enough or those who went over for political reasons or were prisoners who had been promised things. Evenings, the Russians used to broadcast: 'Come over to our side; bring your cutlery, because there's good stuff for you to eat.' And some of them did go over, just like some of theirs came over to our side, but we didn't believe what they said. And I didn't want to stay with the Russians; I didn't want to be a Russian prisoner."[2]

Wounded the Third Time

"January 13, 1945—that's when I got it the third time. Until then, we had been able to hold our positions. The Russians celebrate their Christmas in January, and we had a cease-fire for our Christmas and later for theirs. Someone got Lithuanian ducks for us for Christmas—those were big birds! We were in Tutenfeld—'Totenfeld,' as we used to say [field of death]. It was there in East Prussia that another mine I was defusing exploded. That was my birthday present in 1945, twelve days early!

"I was shipped to Königsberg to recover. While we were in the East Prussian recovery center, a message was broadcast: 'The Russians have us surrounded.' Two days later, we were given the option of staying or trying to make it out on our own. So we tramped from Königsberg to Pillau [now Baltiysk], approximately 45 kilometers. The worst were the Russian Jabos (*Jagdbombers* [fighter bombers]). They threw phosphorus flares and grenades down on us."

The Sinking of the *Wilhelm Gustloff*

"From East Prussia, we headed for Kiel. There was a big KdF [*Kraft durch Freude*] passenger ship that people could take cruises on in the Hitler period. Named after a famous Nazi, Wilhelm Gustloff, it had been transformed into an evacuation ship for the wounded. One of the sailors came over and asked if we wanted to sail home with them instead. We were wearing our winter uniforms, and I think they were shocked at how bloody we looked. I did look pretty bad, because I had been wounded by shrapnel. The sailors hadn't seen so much blood, and they felt sorry for us, so they took us on board. We were taken by dingy to the battle cruiser *Admiral Hipper*, and we set sail, the destroyer *Gross-Deutschland* and the KdF ship *Gustloff* next to us.

"On the second night there was a lot of noise and commotion. They told us to stay calm and that there were military exercises outside. The next day at 6:00, the captain made an announcement: 'The *Gustloff* had been torpedoed last night at twenty-two hours.' Two to three torpedoes had hit the ship, and several minutes later it sank. The spotters reported that only 900 persons had been saved (out of 6,100) in those cold waters. That was January 29, 1945. I remember that like it was yesterday."[3]

Home to Göttern

"In Kiel, we made it to the train station, and there the Guard Dogs [military police] got us. They took us back to camp; split us up; and, depending on what kind of injury we had, shipped us out. I was sent to Bordesholm [near Neumünster], to the military hospital there. You used to be allowed to be shipped back to the hospital near your home, but Himmler did away with that. Then the doctor suggested that my father apply for my release for 'spring planting.' I looked at my arm in the cast and said, 'I will never be able to plant with this arm this spring,' and the doctor grinned and said, 'We know that, but do they know that?' So that's how I got home."

Edgar arrived in Göttern in time to spend Easter Sunday 1945 with his family. Thin and gaunt, his arm in a sling, he was almost unrecognizable to his father, mother, and sister. Overwhelmed with tearful joy, they welcomed him home, counting their blessings to have a son, a brother, return when so many others had been reported killed or missing in action.

Commentary

Edgar's account of his service on the Russian Front marks a shift in the family narrative away from the familiarity of the farmstead and Göttern to a foreign environment made hostile by the presence of the Wehrmacht, to which Edgar was attached. The fact of his being in that place at that time endows the memory of his experiences there with historical import.

It also brings to the surface an underlying point of reference between Edgar and me. We are essentially age-mates as well as combatants and survivors of the same war, albeit on opposing sides, just as Edgar's father and my father were adversaries in World War I, an awesome reminder of how much bellicose twentieth-century history can be concatenated in one relationship. Indeed, when I look at the picture of Edgar in his Wehrmacht uniform, I feel a knot at the pit of my stomach. Infinitely thankful that we can now work together on a project of interest to us both, I find myself wondering, however, how I can truly hear what he is telling me about "his war," mindful as I am of how grateful I was—and still am for my children, my grandchildren, and me—that Edgar's efforts and those of his comrades were unsuccessful.

When I returned stateside after the Aleutian campaign in 1943, I served for a while as a Special Service lecturer at Barksdale Field, in Louisiana, briefing pilots-in-training on the course of the war. I remember with what hope for the future I reported on the success of the Soviet Armies in holding and pushing back the Germans in one appalling battle after another. If I had difficulty understanding what Edgar was saying about those events, it might also be the case that Edgar had a problem in knowing what to say about that war in my presence. But if that is so, I would have had no way of knowing it, for he was as ready to talk about it (albeit probably selectively) as he was about other subjects.

What I find especially interesting in his account are not his references to the fighting as such, which are few, but his observations on the Russian (Edgar never used the term *Soviet*) countryside—the people and their farming practices. It was Edgar's first encounter with a way of life other than his own, and it was the everyday aspects of Russian village life that attracted his farmer's eye.

As a Thuringian farmer, cramped for space, he notes how vast the Ukraine was as he and other soldiers marched through it. Accustomed as they were to their small "towel-size fields," Edgar was amazed at the size of the Russian fields, stretching all the way to the horizon. (Little did Edgar know that he was getting a foretaste of the collectivized agriculture he was to see in his own country within fifteen years.) He notes, too, how rich the Ukrainian soil was compared to his own. And no stones! When he saw the cattle grazing outside, he may well have thought of his father and his dream of pasturing their cows. Edgar's tone in relating what he saw of the Russian countryside communicates envy but not avariciousness.

His remarks about Russians they met along the way are more inquiring than judgmental. He remembers the day he followed the good smell of cooking into someone's house and was rewarded for his boldness with a plate. He recalls the woman who ironed and folded the clothes the soldiers had laid out to dry while they were busy laying mines. "That was kind of dangerous," Edgar comments, "to stay behind as a woman, but we never did anything to them." Edgar observes, too, that there was no sign of electricity in the countryside they marched through.

Edgar notes that Russian children were sent back to Germany: "They worked for us because we had all been called up to battle. . . . We would go from door to door, looking for children." Edgar does not pass judgment on the cruel logic that appropriated Russian children to do the work of German men occupied in destroying the Soviet Union. Perhaps it was thought to be enough that they contravened military regulation by protecting the boy's hiding place rather than shooting his mother on the spot for not divulging a confidence. Is their complicity in the mother's "crime" to be taken as corroborating Edgar's statement, "The people were scared of us, but we never did anything to the civilians"? We do not know the answer; other situations might not have been responded to so benignly.

There is some recognition that the rounding up of Russians to be sent back to Germany to take the place of Edgar and others busy conquering the Soviet Union was connected to the use of prisoners of war (POWs) on Göttern farms. Disingenuously, Edgar asserts that the young people picked up by the Germans were taken "in exchange" for German soldiers unable to work on the farms. In fact, the front and the rear of that theater of operations had become undifferentiated; what Edgar refers to as little more than a reasonable exchange can be viewed as the beginning of a permanent enslavement of Eastern peoples.

Edgar's condemnation is leveled at the German *Schutzstaffel* (SS). He says, "We didn't like them." They did not like the SS because the SS acted without principle, treating fellow Germans as subordinates and shooting their prisoners in direct violation of international law. "But you couldn't say anything," adds Edgar, knowing that if he had, he would have died at the hands of his "comrades-in-arms." In Edgar's case, rather than enticing him to brutalize the enemy in compensation for the oppressive terror that the SS used against him, the criminal actions of the SS worked to weaken his allegiance to the German cause.

The waning months of the war were made more improbable, if not surreal, by having to choose between two remaining fronts on which to do battle: East Prussia or Lithuania. Edgar's choice would seem to have been prosaic rather than ideological. By then, the Russians were occupying most of Lithuania. In East Prussia, Edgar's faith in the German mission evaporated at the sight of young people with only four to six weeks of training being sent to their deaths. "That's when it was over for Hitler," says Edgar. "We didn't believe in it anymore."[4]

Whether the indeterminate "it" refers to Adolf Hitler's cause in general or specifically to the war makes no difference; Edgar had withdrawn his faith. Disillusioned as Edgar apparently was, he did not choose to go over to the Russian side, nor did he wish to become a Russian prisoner. "By the middle of 1944," Edgar says, "the war was already lost." The reason Edgar gives for this opinion is militarily rather than ideologically informed: "Once an army has retreated, it can no longer take a good stand. In spite of the fact that Germans were good soldiers, once he's retreated, he can't take a stand again, maybe for a little bit but not for long; that's how it was in East Prussia."

On January 13, 1945, twelve days before his birthday, Edgar was wounded for the third time and evacuated to Königsberg. As a sensible man would be, he was delighted. "That was," he says, "my birthday present in 1945." His only wish upon recovery was to return once more to Göttern and to his family.

Edgar began the war as a self-proclaimed "fanatic." But the experience of the Eastern Front that drove so many of his comrades to ever-greater degrees of brutalization until the moment when Hitler's death left them disoriented had a different effect on Edgar—that of a man coming of age.[5]

8

~~⌐

American Occupation,
Russian Occupation

Edgar and Wally's Wedding, 1948

The Arrival of the Americans

WHEN EDGAR RETURNED HOME at the end of March 1945, the war had not yet reached Göttern, although American forces were approaching rapidly from the West. Ever since the Rhine had been crossed in March, the Western Allies had raced through Germany toward the Elbe, encountering intermittent resistance. General George Patton's Third Army, driving through Thuringia and aiming for Chemnitz, had Erfurt, Weimar, Magdala, and Göttern in its path. Facing the American forces was the hastily constituted Eleventh Army under the command of General Otto Hitzfeld and later the Twelfth Army led by General Walther Wenck.[1] On April 11, the Third Army, Eightieth Division, 318th Infantry, cleared Erfurt of its stubborn defenders. On that same day, the 319th Infantry demanded the immediate and unconditional surrender of Weimar, while fighter bombers hovered overhead. By late morning, the *Bürgermeister* of Weimar had surrendered the city. That afternoon, the concentration camp at Buchenwald, a few kilometers north of the city, was liberated by units of the Eightieth Infantry Division.

On April 12, Edgar, working high up on the Waltersberge field as best he could with an incapacitated arm alongside his father and the Polish farmhand, Sigmund, saw a line of tanks on the horizon coming from the direction of Mellingen. "What kind of tanks are those?" he remembers asking his father, at that very moment realizing that they were American.

That morning, the First Battalion of the 319th Infantry had received orders to move to Jena by truck. By 15:10, the leading elements had passed through Niedersynderstedt, and at 17:00, the Battalion reached Göttern. There, they detrucked.[2]

Edwin, Edgar, and Sigmund put down their tools and hurried home, concerned for the women. By the time they had returned to the village, American soldiers were everywhere. "They didn't do anything to us. No one tried to resist," says Edgar. The commanding officer requisitioned the Schorcht house to billet his staff. He took over the *gute Stube* as his command post, where only a few days before an SS officer, in command of a Buchenwald "death march" that passed through Göttern, had taken his midday meal.

Edwin moved the family across the courtyard to the quarters above the washroom, where they had been accommodating evacuees. There was a knock on the door, and it was a young American officer who told them, in near-perfect German, that he was staying in their house and wanted to give them something for their troubles. His grandparents, he explained, were German. He presented Elly with a box of cookies and Edgar with a pack of cigarettes. As Edgar says, "It didn't make any difference that I didn't smoke. What I saw then in comparison to our soldiers left an impression on me." Irmgard was less grateful than the others. She says, "My mother befriended them right away and accepted their gifts, but I couldn't. I didn't take anything from them. One night the Czechs in the U.S. Army—there were several of them here—tried to come up and 'visit' us, but our Poles, the people who worked for us, protected us from them."

As seemingly friendly as the Americans were, the Schorchts were uneasy in regard to what lay in store for them, especially when a house-to-house search was begun to round up anyone who had served in the Wehrmacht. Arriving at the Schorcht house, they called for Edgar, already knowing he was there. "Come on down," they ordered. "We have to take you with us." Even before Edgar had time to appear, the young American officer, who spoke German so well, said to his commanding officer, "Look at his arm, sir; he won't do anything." The commanding officer nodded in approval, allowing Edgar to remain at home under house arrest.

Two days later, it was announced that every man of military age had to report to the town hall. Edgar says, "My father was concerned about what might happen there. I put on the parts of my uniform that remained, for we knew it would always go worse for men who weren't soldiers [not in uniform]. Otto Hartwig went with me and another guy from Philippsthal [west of Eisenach]—Ernst Bonnewitz was his name. He had been visiting his aunt here. He had been in the quartermaster general corps and so didn't know anything about war. Anyway, the Americans had been in his aunt's house and saw a picture of him in uniform. I remember he was wearing a sweater when I saw him in the town hall. That was the end of him! They took him away on the spot. At the town hall, they removed everything we had with us, from pocketknives to wallets, and then we went into detention."

Everyone who had been rounded up was loaded into a truck and driven to Nohra, where each prisoner was inspected for the tattoo on his underarm that would identify him as a member of the SS. There Edgar observed the emaciated former inmates of Buchenwald identify who, if any, among the German soldiers had been their guards and torturers. Edgar says, "We were scared of what the Americans might do to us. Had we been prisoners of the Germans, we would have been slaughtered. Hitler or Himmler—they took no prisoners. What we experienced in the [American] camps was nothing [compared] to what we had done." (The "we" in this assertion is not specified.)

From Nohra, they were driven to Gera. At Gera, the convoy stopped. "There," Edgar says, "an American soldier, probably a Jew, searched us. We had to empty our pockets. One guy had his house key taken from him and said, 'That's my house key!' and the American soldier said, 'House key! What will you need a house key for? We are handing you over to the Russians.' We were really scared. No one wanted to be handed over to the Russians. When we got back in the trucks again, we noticed that all the gas canisters were empty, and we said to each other, 'They are not going to drive us to Russia with empty gas cans.' We felt better."

Edgar was amazed at how much time was spent driving them around in trucks. If one of the trucks broke down, it would often be an officer, rather than an enlisted man, who would get underneath the truck to fix it. Edgar says, "All it took was one whistle to get the convoy rolling; you can't imagine what it would have taken to get us [Germans] ready to move, how many officers and assistants would have had to give orders that would have had to be previously approved. The Americans, one whistle—that was enough."

As for food, Edgar says, "What we got from the Americans to eat was good, but you have to understand—Germans need only bread, potatoes, and meat to survive. Everything we got [from the Americans] was in cans: meat in cans, potatoes in cans, and no bread!"

From Gera, the prisoners were driven back toward the West to Bad Hersfeld, north of Fulda; and then later to Heidenheim, north of Mainz; and finally to Bad Kreuznach, just south of Mainz. In the course of a few weeks, thanks to the American Army, Edgar saw more of Germany than he had in his entire life. He and his fellow prisoners were detained in Bad Kreuznach for thirteen weeks. He says, "I knew that it was going to be bad. The commander of our camp was Jewish. We knew, of course, what Hitler had done to the Jews, so we expected the worst. The good thing about Bad Kreuznach was that we were fenced in under cherry trees. When the cherries ripened, we picked them and divided them evenly. It was about a month after the cherries ripened that we were released."

By the time Edgar returned to Göttern, the village was in the hands of the Russians.

The Yalta Agreement

It had been agreed among the Allies at Yalta on February 11, 1945, and previously by the European Advisory Commission on September 11, 1944, that Thuringia as well as the western portion of Saxony would be included within the Soviet sphere of occupied Germany.[3] When hostilities ended in May 1945, the fortunes of war had been such that the final assault on Berlin had precluded the Russians from crossing the Elbe to the south. The Anglo-American forces had already swept with relative ease through most of central Germany and were poised to cross the Elbe. General Dwight Eisenhower restrained them from doing so. When the Americans withdrew from Thuringia in the first week of July 1945, allowing the Russians to enter triumphantly, they did so in accordance with the prior overall strategic agreement among the Allies. Even now, people in Göttern speak with bitterness of what they regard as a deal made between the Americans and the Russians at their expense. Wally's voice hardens when she says with indignation, "We were traded away for Berlin, you know. The Americans might have stayed here, but instead they traded us away for Berlin."

Edgar Returns from American Internment

The family had heard nothing from Edgar since he had been taken away by the Americans in April. It was hard keeping their spirits up in the face of his absence. Irmgard recalls the dream she had foretelling Edgar's return. She says, "I told them that Edgar would come home safely. I saw him in a long row of people. I had to soothe my mother somehow. Then one day we were mowing clover up by the Autobahn, and soldier transports were driving by. My father found a four-leaf clover. My father was always finding them; he used to send them to me. Anyway, he found one and said, 'If I could wish for anything, I wish Edgar would be there when we come home tonight.' We came home through the back entrance. One of the daughters from the East Prussian family we had taken in came running and said, 'Irmgard, Irmgard, we have a visitor!' I asked, 'Who? You can tell me.' She said, 'Your brother is here!' Just a few hours before he had made our wish come true; we were overjoyed.

"Afterward we saw transports. The Americans were leaving. We knew the Russians would come. We saw people shoving pushcarts—refugees trying to get back home [der Treck]. We didn't have beds for all the people sleeping in our house. My father put down some straw on the barn floor, so they could sleep on that. My mother gave them a little something to eat in the evening and some coffee in the morning, with a little something for the children. During the day they would mostly sit around in the courtyard, clutching their things and staring off into space. Then they would continue on. We had new people almost every night."

The Americans Leave; the Russians Arrive

Edgar returned to Göttern with trepidation. His concern had to do with Edwin and Irmgard, who as Nazi Party members faced arrest and incarceration, if not worse. Hard times were in store for everyone—especially, it was feared, the women. Some gave way to panic when they thought about what might befall them at the hands of Russian soldiers. Wally spoke for many in saying, "When we were told the Americans would leave and the Russians would come in, we were frightened for our very lives." Apprehensive as they were at the prospect, it was the first sight of the Russian ponies pulling carts, rather than the large tanks, that remains in the mind's eye of the Schorchts.

Edwin Is Interned by the Russians; Irmgard Goes into Hiding

When Communists in Magdala made themselves known to the Russians, they also identified Nazi Party members in the area. Edwin was commanded to present himself to be transported to Milda for detention. He dutifully did so. Elly was distraught, but she did not try to prevent him from going for fear of what might befall him if he did not follow orders. Irmgard was supposed to go as well; however, she was determined not to. She says, "My mother kept urging me to go, and I told her, 'I'm no longer a minor. You are not responsible, and you don't know where I am.' The mayor's daughter came to me and asked if I was going to go, and I said, 'No, I am not going to go, and I advise you not to go as well.' But she went anyway, and the Russians took her and they surely raped her—she was the only woman among all those men! The next day, they fetched a doctor. That could have happened to me—that's why I didn't go. That day I said, 'You see now, Mother?' and she replied, 'You were right.' She was so scared; I told them they needn't fear for my safety. No matter what my parents said, there was no chance that I would be going with the Russians. I hid myself in a special place in the barn."

According to Edgar, these frightening events happened when the Russians first came. "They were dangerous then," he says. Irmgard vividly recalls one experience: "One night they suddenly appeared. They were in a sorry state, and we were still afraid of them. Our mother was so delicate, but she was courageous too. She heard something in the courtyard and peeked out of the window and she saw them back there—Russians. What did she do?

She had her slippers on, and in her nightgown she went out there. They had guns. And she said, 'You boys had better get out of here,' or something like that. They understood that they were unwelcome and left. I remember Edgar and me at the window, screaming for help, and people came with pitchforks or whatever. But they were gone by then. You couldn't do anything with them around. They'd pick your potatoes at night, and their little horses would clean out your fields."

Edgar clearly remembers the anxiety he felt for his father, detained by the Russians in Milda. He believed that the Russians would have to believe his father when he told them that he was not a convinced Nazi but had joined the party only when he was persuaded that it was for his children's future interests to do so. Whether the authorities were moved by that explanation or influenced by Edwin's upright appearance, he was in any event released in a matter of days. Edgar remembers the joy he experienced watching out the window for him, as he did every evening, to see him, one moonlit night, cross the bridge on his way home.

A Training Farm

By the fall of 1945, Edgar's arm had healed sufficiently that he was able to work beside his father full-time. The Polish workers had left. The Americans had taken Sigmund first to Erfurt, then later to West Germany. Edgar says, "When the Americans came, it was like a little revolution for [the Polish POWs]. They thought, 'Now we are free.'" Eventually Sigmund found his way home to Poland and to his family.

Even before the war ended, Edwin's plan to have the farm registered for apprentice training for males and females had been approved. Now that the Poles had left with the Americans, it was all the more important to have additional help. By the time the Soviets arrived, there were two apprentices on the farm. Like almost everyone else, they went into hiding at the first sight of the Russian soldiers. Later, when some degree of normalcy had returned, there were at any one time two young men and one young woman in training under Edwin's and Elly's supervision, assigned by the Agricultural Bureau. Edgar says, "The drawback to the program was that the apprentices came to the farm with little experience, most of them being fifteen to sixteen years of age. After two years' training, they would leave just as they were becoming skilled." In addition, there was one hired man to help with the animals.

Irmgard Tries Again

On the completion of her own apprenticeship on the farm in Apolda, Irmgard had acquired the qualifications to train agricultural apprentices. She traveled to training farms on her motorcycle, disassembling parts of it every evening to deter the Russians from requisitioning it, to demonstrate to the apprentices how to clean cows and horses (she had learned from her father the proper way to clean an animal). She also gave instructions in cooking and sewing. Edwin continued to see to it that the cows were washed every Saturday, their tails scrubbed with soapy water, and the stalls whitewashed twice a year: on *Kirmes* in the fall and on Ascension Day in the spring.

When Irmgard was home, she helped her mother with the milking. While not as adept a milker as Elly, she prided herself on how fast she could fill a pail. One day she was kicked by a cow, suffering a cracked vertebra. Although Irmgard healed well, the doctors told her that she should not go back into agricultural work. She applied for admission to the *Arbeiter-und-Bauern-Fakultät* (worker and farmer faculty) program in Jena, hoping to make up for not having attended *Gymnasium*. Her dream had been to study medicine. Irmgard says, "I like to help people. I was the one they called when someone in the village was sick. Before the doctor had a chance to arrive, they got me. That was the case with our neighbor who tried to hang himself. He had become depressed. I was the only one around who had a premonition of what was going to happen. I could intuit those things. That is why they came running for me. So I wanted to register for the medical exam but had to have the permission of the local authorities to do so. My father said, 'What? You want to go to those people [Communists] who want to take everything away from us?' So then it was all over. As I look back, I see now that I was always the one who made the sacrifices. I know that now. . . . I should have been more self-interested. I let that happen, even though I would like to have gone to the university."

Edgar Takes to Dancing

By 1946, Edgar had regained a good part of the weight he had lost, although his cheeks remained sunken. Working side by side with his father from dawn to dusk, he began to feel more like himself at his best. Slowly, memories of the Russian Front began to dim. More and more, he felt like going out with his old friends, even though he did not care for alcohol. He kept in mind what his father told him: "My boy, remember that the ox drinks only as much as he needs. What about man? He drinks more. You must not put yourself below the ox." Edgar says, "I took that to heart. We were a hearty group; we did everything. My trick was that I could be happy drinking soda water. Others had to drink beer or a bottle of wine, but not me. That's an art."

What Edgar discovered he wanted to do more than most of his friends was to go dancing. Both he and his sister had learned the rudiments of dancing from their mother. Elly enjoyed dancing and hoped her children would too. The opportunity to learn further came one day in 1940 when Herr Harry Stroh, the local dancing teacher, called at the Schorchts' house, hoping to recruit additional students. Edgar somewhat reluctantly agreed to attend. The lessons were held once a week in Magdala. There were always more females than males, because by 1940, some of the young men had already been called up.

In the beginning, Edgar did not enjoy the lessons. The teacher insisted on the students' learning the steps first before pairing up. Edgar had come to meet girls, so as soon as they began dancing in pairs, his spirits lifted. Sitting in a circle around the hall, the mothers kept a close watch on who was dancing with whom. As Wally later said, "It was nice then, before the war. There was a real cohesiveness."

Six years later, in 1946, Edgar discovered that he had lost none of his agility on his feet. With his jet-black hair combed back, he knew that he cut a good figure on the dance floor. Whenever he heard of a dance in one of the nearby towns, he made a point of attending. It was on just such an occasion in neighboring Bucha, at a Ladies' Ball, that he met Wally.

Wally Hünniger

Wally was born in 1926 in Bucha, about 4 kilometers southeast of Göttern. She was also an enthusiastic dancer, having taken lessons from the same Herr Stroh. Before long, Edgar found himself paying Wally a visit once a week. They got along well, she laughing at his jokes, he beguiled by her beauty. They both knew each other to be serious and hard workers. Before many weeks had passed, Edgar asked permission of Wally's father to hold his daughter's hand. In 1947, they became engaged, Edgar making a formal call on Herr and Frau Hünniger to ask for their daughter's hand in marriage. Edgar enjoyed the occasion, for he had come to know Herr Hünniger as a friend. The problem, however, was with Wally's grandfather, who lived in the Hünniger household. A representative of the Raiffeisen Bank and the owner of a fertilizer business, he was always referred to as "Herr Minister" because he was considered to be smart. The grandfather was set against the engagement, as he had someone else in mind for Wally, a young official in the bank, who would take her away from the rigors and smells of farm life. "But he won," says Wally with a laugh, pointing at Edgar. "It was hard, but I did it," replies Edgar with a broad grin.

Like Edwin, Wally's father, Erich Hünniger, was a farmer, though on a somewhat smaller scale—18 hectares as opposed to Edwin's 28—but still large enough to support two horses, cows, and a household consisting of Wally's parents, her grandfather and grandmother, her uncle, a sister, a brother, and a servant girl.

Wally always strove to do her best. She took her eight years in the Bucha school (one teacher for sixty-eight students) seriously. She was confirmed following two years of religious instruction. For the day of the exam, she wore a new rust-brown dress, one that remained her favorite for years. The village minister had already been called up for military service, so another minister was brought in from nearby Leutra. He helped the applicants by telling them beforehand which questions he would ask. The applicants, eight girls and four boys, all passed without embarrassment.

On Palm Sunday, Wally wore a new black dress, especially sewn for the occasion, her hair done in two braids down her back, and around her neck a gold pendant. Her father and grandfather wore starched wing collars, striped ties, and dark suits. The two godmothers and godfathers came to the house bearing gifts of money. Some of the other guests, close relatives, brought small presents, soap or cologne. For lunch, there was soup, carp, and three kinds of meat. After the ceremony at the church, the guests were invited to the house again for coffee and cake and the evening meal. As the guests left the house, they were once more given packages of cake to take home.

The following day, Monday, freely taken off by farmers who had worked the full week before, saw the festivities continue. The godparents, relatives, and friends who had brought gifts came to the Hünniger house again for coffee and cake and later an evening meal. On leaving the house, they were given packages of cake to take home, some now for the third time.

Wally looks back on those days with wonderment at the amount of time she, her mother, and other women spent baking cake and bread every week in wood-fired ovens, the yeast for the bread being made from wheat as well as from potato. The variety of cakes was myriad: bacon cake, onion cake, quark (curd cheese) cake, poppy-seed cake, and fruitcake, among many others. They continued baking that way until the mid-1960s.

Wally finished school in 1941, the year her father, her brother, and her uncle left for the war. Hardly had the men of the family gone when the *Schutzstaffel* (SS) appeared and appropriated the tractor that the family had purchased the year before. Wally was left alone with her mother to operate the farm with the help of two POWs: a Pole and a Frenchman. Wally's sister, younger by four years, was still in school.

It seemed as if the war years would never end. Even after the surrender in 1945, none of the men came directly home. Wally's father had returned shortly before the end of the war, but he was called up again for the *Volkssturm* (German Territorial Army) and was taken prisoner by the Russians. He did not return to Bucha until 1946. Days and weeks turned into months, and still no word of Wally's brother, Reinhardt. The last report of his whereabouts came from somewhere in Romania, through which the Germans had retreated in January 1945. Wally says with a profound sadness,

"The soothsayers always said, 'The brother will come home. He's alive; he's coming home.' They always saw it in the cards. We were not the only ones who did that [had the cards read]; lots of people did. Lots of people had hopes for men who never came back."

Preparations for Edgar and Wally's Wedding

The day of Edgar and Wally's wedding was fixed for September 19, 1948, in the church in Bucha. There was much to attend to, especially as everything was in short supply after the war. Both families had been saving up flour, sugar, and butter for months for all the baking that had to be done. Many of the ingredients were obtained through barter. The sugar coupons were combined, and a little sugar was set aside from each coupon every week. Homemade wine was produced in great quantities, from apples, pears, gooseberries, currants, and blackthorns. The amount of hoarded sugar added enhanced its potency. Schnapps was made secretly in the steamer used to process potatoes for the pigs. There were also two cases of superior Mosel wine sent during the war for safekeeping in Göttern by Elly's brother Walter, "the Nazi." Irmgard helped her father bury it in the barn until the great day arrived. Edgar had already bought two wedding bands in Jena from the same jeweler his father had gone to in 1919 for Elly's engagement ring.

In more normal times, Wally's family would have provided her with a larger trousseau than they could manage in those lean years. To obtain even the minimal amount of sheets, tablecloths, and towels, Wally's mother was constrained to trade for them with people from Jena who were eager to procure firewood and sausage. As for the furniture, she provided bedroom and dining-room furniture as well as the large table that is used in Wally's kitchen today. Fortunately, a distant relative, a woodworker in Jena, bereft of wood, made the pieces with ash provided by Wally's father.

The Wedding

The morning of September 19, 1948, broke clear and chilly. Edgar and Wally made a point of being at the registry even before it opened at 9:00, so they would be sure to be on time. They were accompanied by two witnesses. The civil ceremony did not take long. Afterward, they walked to the Hünniger farm on the edge of town to wait for the guests. Wally's long white dress was the subject of considerable speculation. Above all, people wondered where such lovely material could have been obtained in such difficult times. The truth of the matter was that Wally and her mother had traded feather bedding and a considerable amount of well-cured bacon to acquire it.

Edgar looked handsome in his white bow tie and his new dark suit made by Herr Collitz of Magdala, the same tailor who had sewn his father's suit. People remarked how thin the bride and groom appeared. Now, when Edgar looks at the picture taken of them that day, commenting on how thin everyone appears, he says, "The wind still blew through our cheeks then."

When the guests had all arrived, the procession was formed to walk to the small baroque church with the massive spire, shingled in black slate. As they approached the church, children in white spread flower petals along the nave up to the altar. Behind them came the bearers of the two embroidered cushions upon which Edgar and Wally were to kneel. Then the relatives entered two by two, first Wally's parents, followed by Edgar's.

When the final vows were exchanged and the wedding party left the church, there appeared, to everyone's surprise and delight, a chimney sweep. Everyone knows that a chimney sweep brings good luck when he comes by just after the service, especially when he kisses the bride. Nearby, the children were waiting for Edgar to throw them money. Farther along the road to the Hünniger farm, children had stretched a rope across the way. Additional "ransom" money had to be paid for the rope to be lowered.

At the door of the house, the bride and groom were offered the traditional bread and salt. The festivities went on for several hours, with course after course spread before the guests. The repast started with clear soup with small dumplings; followed by three courses of meat, beef, pork, and schnitzel; and finally a good dessert, lemon cream. All during the meal, the bride and groom sat close to each other so that they could eat from the single dish that custom allotted to them.

It was said more than once that nearly everyone at the wedding reception was a farmer or a farmer's wife, with hearty appetites that could do justice to two feasts in one day. It was late when the guests finally left, tired from such a full day of festivities.

For Wally, the celebration marked the beginning of her new life, the first night in her new home. She knew it would take patience to accustom herself to living with her parents-in-law. Edwin and Elly Schorcht were still the master and the mistress of the household; it would be some years before she and Edgar took their place.

Meanwhile, there was work to be done. There were the animals to be fed and milked early next morning. They would need to be up and about by 5:00. When the final glass of wine and coffee had been drunk and the last guest had departed, Edgar and Wally left for bed, happy but exhausted, only to find that their bed was not there. In keeping with tradition, some of the young people had dismantled it and hidden it elsewhere in the village!

Commentary

Between the time that Edgar returned home from the war and the arrival of the Americans, an experience involving the inmates of Buchenwald, the SS, and Edgar occurred in Göttern, leaving an indelible impression on Edgar's mind.

Some years later, in his capacity as mayor of Göttern, Edgar received a letter, written on April 7, 1985, by members of the Young Pioneers and the *Freie Deutsche Jugend* (FDJ) of the high school in nearby Milda asking him to provide information about a trek of inmates from Buchenwald who might have passed through Göttern. As part of their training in anti-Fascism, the students, taking part in a project named Red Triangle, focused on researching the marching routes of the inmates who were forced to leave the KZ Buchenwald on "death marches" (*Todesmarsch*) in April 1945.

On April 7, 1945, more than three thousand Jewish prisoners were sent toward the Flossenbürg concentration camp by foot. The trek proceeded first in a southeasterly direction to Gelmeroda (made famous by Lyonel Feininger's 1913 sketch of the Gelmeroda church), then to Bad Berka through the woods and side roads toward Orlamünde. This trek, or a smaller part of it, passed through Göttern.[4]

Edgar remembers their arrival in Göttern well. He says, "I had been fighting for four years, and we had never seen anything like it in all that time. Right in our living room in the *gute Stube* there was . . . an SS officer . . . here to eat dinner. No one dared to go in there except my mother. But the others, the guards and the rest, they didn't get anything to eat. They [the prisoners] were packed in the barn [the former Dobermann estate] the way you pack sardines in a can. Wagner [who leased the Dobermann estate and was a friend of Edgar's] came over and asked me if I could lead the horse, because he wanted to get his wagon out. So we got it out backward with the horse, and he was steering from behind the way we did it back then, and I could see the 'Buchenwalders' looking at the fodder beets. A couple came over and asked if they could eat some of those beets. And Wagner said, 'Take them, take them.' And as they bit into the beets, the SS came and started laying into them with sticks. I said, 'Hey [*in a scolding tone*], what's wrong with you boys?' and Wagner said, 'Be quiet, or they'll put you in with the rest.'

"There were a couple hundred," Edgar recounts. "They were divided into two groups. They stayed overnight, and then the next day they headed up for the Autobahn. Five of them couldn't go any farther; they were shot, and they were buried up there—the prisoners had to bury them, of course—in a very shallow grave. When that was all over, a few [Communist] Party members disinterred them and reburied them in the cemetery."

In the close to his letter to the school, Edgar writes, "At the beginning of my term as mayor, a proper gravestone was put in place. For us, it is a special responsibility to take care of this resting place of five unknown fighters for peace. . . . We would like to show you, dear Pioneer and FDJ-members, the memorial site, which is a protected landmark."[5]

Edgar makes reference to the incident on three occasions in the course of the interviews. Each account is essentially the same, the emphasis being on his intercession on behalf of the inmates. According to Daniel Goldhagen, the response of the German populace to the columns of emaciated specimens of humanity moving silently through their villages and towns was, more often than not, hostile.[6] Assuming the truthfulness of Edgar's account, his solicitation for the inmate's welfare, while not unique, was singular. It would seem to absolve Edgar of Goldhagen's judgment that "'ordinary Germans' were animated by a particular *type* of antisemitism that led them to conclude that the *Jews ought to die*."[7] Indeed, from Edgar's perspective, he was in the moral right, as if his retelling of the story suggests that what he said in Wagner's barn compensated for his failure, or at least his inability, to intervene on behalf of the Red Army prisoners captured by him and his comrades in 1943 and 1944, who were then shot by the SS. The family had acceded to authority by putting their best parlor at the disposal of the SS officer. To submit to the intentions of authority was a normal expectation for the Schorchts. To contravene that commitment was, in Edgar's opinion, a principled act.

What is worthy of reflection is the Götterdämmerung atmosphere that must have prevailed in and around Göttern at the moment at which the drama recounted here was merely one act: the bombing of surrounding cities, American troops approaching, soldiers returning, refugees streaming in and through the village, the evacuation of Buchenwald, and the knowledge that the "inhuman hordes" somewhere to the east were on the march westward.

For the people of Göttern, who had lived in relative tranquillity during the years of National Socialism, the poisonous fruits of Hitler's ideology had been brought full circle to their doorsteps. They finally had occasion to know the truth about Buchenwald.

9

Land Reform, 1945–1949

OR THE SCHORCHTS, as for all rural people living in the Soviet zone, the most far-reaching event in the immediate postwar period, other than the arrival of the Soviet Army itself on German soil, was the imposition of the land-reform program (*Bodenreform*). From 1945 to 1949, the energy and attention of much of the population was focused on the fundamental effort to alter the nature of land control and thus the distribution of power in eastern Germany.

At the war's end in May 1945, the Allies were in agreement that fundamental changes needed to occur in the structure of German society to prevent the recurrence of its militant, destructive tendencies. First and foremost was the perceived need to root out the reactionary Junker class, with its hold on German rural life, its inordinate influence on the course of German politics, and its fostering of Fascism. The vast latifundia east of the Elbe had to be eliminated, and the land used to provide farms for countless refugees, displaced farmers, farm workers, and dispossessed urban workers. The rural areas of East Germany were to provide the initial fulcrum for fundamental change in German society.[1]

On September 10, 1945, the provisional government of Thuringia approved the KPD (German Communist Party) proposal for land reform, describing it as "an urgent, national, economic, and social necessity."[2] It called for expropriation of all land belonging to war criminals (leading Nazis or Nazi supporters), estates of more than 100 hectares, and land belonging to the state or to its institutions. These holdings, including buildings, were to be apportioned among peasants with fewer than 5 hectares of land, landless peasants, agricultural wage workers, small tenants, the dispossessed, and refugees.[3] Each portion of land was allotted at a cost equivalent to one year's harvest, to be paid either in rent or in kind over the next ten to twenty years.[4] In each village, a five- to seven-member commission—elected by an assembly and comprising peasants with fewer than 5 hectares of land—would submit two lists to the German administrative authorities of its district. The first list would specify the properties to be confiscated, and the second would identify the eligible recipients of the land. It was to be the work of the local commission and the peasants to confiscate the land, evict the former owners, and distribute the property.[5]

The "Large Farmers" under Siege

As the owner of a 29-hectare farm, Edwin Schorcht had no reason to fear expropriation. Yet the threatening tones in which land-reform goals were couched were unsettling. There were frequent reminders that "large landowners" were capitalists and enemies of the people, if not war criminals. "The land reform was a terrible time for the old farmers of the region," says Edgar. "All those who had over 20 hectares were notoriously known as 'Grossbauern' [big peasants]. We were enemies at first. Although it wasn't all that bad in Göttern, in general that's the way it was. Whoever had more than 50 hectares, if he didn't watch out or couldn't make his quotas, it was all taken away, just like that," says Edgar, with a snap of his fingers.

The Schorchts understood that the wisest course for them was to keep quiet, remain strong, make their quotas, and have as little to do with the land reform as possible. The fact that the name Schorcht appears nowhere in the land-reform files pertaining to Göttern is a reflection of their success. Edgar says, "We couldn't look into what the land reform was doing too much; only those who received something from it could. But the others . . . those of us who were known as 'Grossbauern,' we had to be careful."

Land Reform Begins in Göttern

In keeping with the KPD's sense of urgency, the first step toward implementing land reform in Göttern was taken fifteen days after the program had been approved for Thuringia. On September 25, 1945, the then mayor of Göttern, Gerhard Rothe (himself a small farmer with little land), submitted the names of six members of the newly nominated local land commission.[6]

On September 29, Chairman Kraft informed the county (Kreis) commission that there was no estate of 100 hectares or more in Göttern to be divided up.[7] A demand for land, however, was coming from a number of sources, including the evacuees from the East. On September 27, the county commission in Weimar received a letter from Lothar Becker, formerly from Guttstadt, East Prussia, asking for farmland. Sometime after his arrival in Göttern, Becker had married the daughter of the mason Walter Böttner, like him a member of the local commission. The following day, there arrived at the county commission a letter from Lothar's father, Alfons Becker, also asking for land.[8]

The largest property in Göttern, and as such the only possible source of land for subdivision, was the former Dobermann estate, located diagonally across from the Schorcht farmstead. It consisted of 54.83 hectares, a large dwelling and barn, stables, and numerous outbuildings. The Dobermanns had sold the property, land, and buildings to the state between the wars. As a state domain, it had been subsequently leased, first to the Fechte family and then to the Wagners (into whose barn the survivors of the death march had been forced to enter by the *Schutzstaffel* [SS] in April 1945). The Dobermanns left the area. According to Edgar, the young Ernst Dobermann had wanted to marry Elly Michel, Edgar's mother, but never returned from World War I.

Edgar has fond memories of the Wagners. He says, "Yes, they were fine people, the Wagners. When I was in the army in Weissenfels at the beginning of the war, the Wagners were the only ones with a phone other than the mayor, but we couldn't disturb him—he was so unfriendly. But when I called the Wagners, I could hear Frau Wagner with her wooden shoes, bearing my message out the door. That's just the way it was; that's how those people were, helpful. . . . When we didn't have a tractor back then, and we had fallen behind, he just said [*Edgar imitating*], 'Ah, take my tractor . . . the large Bulldog.' And we just filled it up with gas afterward; we didn't have to pay anything. . . . The times were just very different back then, just before the war."

On October 10, 1945, the police took Wagner into custody on the basis of resisting entry, leaving his wife alone with a newborn baby and four other children.[9] Walter Schulze was named administrator of the estate, later to be replaced by Michael Winter. The local commission faced a dilemma: to act decisively on the Dobermann estate or not to act at all. By the first week in October, the local commission made the decision to divide the property of the state domain to meet the needs of the landless peasants, those with little land, and the several refugees from the East who had settled in Göttern.[10]

There was considerable sentiment in Göttern, however, that Wagner was being unfairly treated. Both the mayor and Schulze signed a letter to the police authorities in Weimar on October 11 to the effect that because Herr Wagner enjoyed the highest reputation in Göttern, had been helpful to the town inhabitants in every way, and was not a Nazi criminal (albeit a party member since 1936), and because the state domain leased by Wagner lay well below 100 hectares—54.83, to be exact—he should be released.[11]

On October 10, the day Wagner was taken into custody, Herr Ritter from the district commission in Weimar paid a visit to Göttern and the state domain to take matters in hand. He submitted a report to the commission, stating that while the lessee Wagner was imprisoned, Frau Wagner was still living on the estate with her five children. He requested that she leave Göttern within forty-eight hours. All the furniture would have to be inventoried. He noted that Winter, who had been put in charge as administrator, had shown no particular interest in his assigned task. Accordingly, Winter was told that as of the next day, a trustee would assume control of the estate and that Winter would have to hand over everything in an orderly fashion. Most regrettably, little had been done by way of dividing up the property. The local commission was informed that it should start surveying the land immediately. It was made clear that as there were only some 54 hectares to distribute, no former Nazis would be eligible.[12]

With Ritter's departure, however, matters did not proceed as he had specified. On October 15, Schulze informed the district commission that, lacking a consensus, it had been impossible to proceed to the division of the estate on the requested date. Further, he asked the commission to allow Wagner to retain a *Restgut* (portion of the estate) to secure the livelihood of the four families depending on him.[13]

Wagner's advocates won additional support on October 18, when three of Wagner's farm workers penciled a note to the district commission, asserting that according to the provisional division of the estate, only farmers were allotted land, but nothing had been given to farm workers. Furthermore, they asserted that the vice chairman of the commission, Günter Müller, was known to have been a Nazi Party member. The three signatories closed by saying, "The undersigned ask urgently for a strict inquiry, because they feel themselves to be totally discriminated against."[14]

Accusing commission members of having been Nazi Party members was an effective way of deterring the work of the commission. Müller, who had applied for 2 hectares of land and had been appointed to the commission by the mayor, was now threatened with dismissal from the commission and rejection of his request for land on the grounds of having been a National Socialist German Workers' Party (NSDAP) member in Weimar.

On October 28, Müller wrote to the district commission, objecting to the measures to be taken against him. In his defense, Müller claimed that he was coerced into becoming a party member in 1933. When he became aware that the NSDAP was not a socialist workers' party, he withdrew his membership in April 1935. Müller closed by saying that he always was and still remained a social democrat. Attached to Müller's letter was a note of support from Herr Freitag of the county commission, affirming that Müller's early departure from the Nazi Party proved that he had been an anti-Fascist from that time. He added that the 2 hectares of land should be assigned to him as long as no other anti-Fascists applied.[15] In his capacity as county commissioner, Herr Freitag saw it as his duty to support the local commission against the supporters of Wagner. Herr Freitag reported regularly to First Lieutenant Subkov, a member of the Soviet Military Administration for Thuringia in Weimar.

Müller's letter and Herr Freitag's note were, however, to no avail. At a meeting of the general assembly on October 27, the day before Müller had written his letter, members had already agreed to dismiss him from the commission. His place was taken by Kurt Vogel, one of the farm workers who had accused Müller of being an NSDAP member. The assembly also voted unanimously to empower Alfons Becker and Walter Schulze to survey the lots to be distributed in the name of the local commission. It was then seventeen days after Ritter had said that the surveying was to commence immediately.[16]

This was a difficult time for the Schorchts, with accusations being leveled against large peasants, threatening that they would have their comeuppance one day. The Schorchts wanted to keep out of the way as much as possible, but that was not easy when friends and neighbors, like the Wagners, were in dire straits.

Edgar remembers working in the mill with Erich Ludwig (a fellow *Grossbauer*, living just behind the Wagners' place) the day that Wagner's wife, Anna Laura, came running into the village to say that they were about to take her furniture. In the dark of night, Edgar enlisted the help of others to save most of it. Even so, they knew they were being observed. Edgar says, "We borrowed a tractor from the Ludwigs. Wagner himself had a '28 Deutz, but we couldn't use that because they had already laid their hands on it. We retrieved two wagonloads of her furniture, which was sent to relatives in Altenburg. The people who were acting up then were the people who had something to gain from this stuff. No one could stop them. They did a job on those people, just like they did with the Jews, just because these people owned land; that's how fast things like that can happen."

On November 6, a new election was held for the local commission. Becker was unanimously elected chairman, and Georg Kraft was made deputy.[17] Becker, who was not afraid to speak up, was determined to make a way for himself in Göttern and help his son Lothar as best he could along the way. Becker had a keen sense of those he believed were hindering the work of the local commission.

On November 11, Becker wrote a letter from his sick bed to the county commission in Weimar, in which he alluded to "dark forces at work" in Göttern that "want to hinder in every way possible the good work of the land reform." He listed those he believed to be the principal culprits, including Wagner and his wife. "So today they brought out another two wagonloads [of furniture]," he wrote, "to Erich Ludwig, without saying anything to me. They should be out tilling the fields instead." Becker continued to list misdeeds perpetrated: machinery and equipment left out in the fields, turnip rows not covered, and, worst of all, "They lock up the desk to prevent me from looking at the books." And more complaints: "Schulze goes to the estate with empty pockets and leaves the farm fully loaded. . . . You rarely see him during the day, only in the evenings or at night." Becker made a motion at the end of the letter to arrest Schulze as well as "his accomplice Georg Kraft and to search their houses." Becker warned the recipient of the letter to handle it confidentially because "they are foxes." In a postscript, he wrote, "This and other injustices are happening in Göttern. It is all aimed against the poor and the poorest; there must be an end to this!"[18]

In late fall 1945, the division of the state domain finally occurred. Of the 54.83 hectares, 13.25 were allocated to farmers without land; 24.25 were divided among six farmers with small amounts of land; and 17.33 were allocated to refugees, including Lothar Becker and his father, to small leaseholders for gardening, and to the village of Göttern itself, to dispose of as it saw fit.[19] Following a good deal of discussion, the apportioning of the livestock took place on November 3, a total of fifty-two animals: six horses, twenty-three head of cattle, eleven pigs, and twelve sheep.[20] These animals had been the personal property of the Wagners, unlike the state land, which they had leased.

To complete the process of allocation, recipients had to submit a declaration that they had not been members of the NSDAP, the *Wehrmacht*, the Waffen-SS, the *Sturmabteilung* (SA), the National Socialist Motor Corps (NSKK), or the National Socialist Flyers Corps (NSFK). If they had been, they had to state their period of membership and the rank or office they held.[21] Of the declarations, most were handwritten by Mayor Rothe on his stationery and then signed by the recipient. Only the Beckers used the official forms. On November 22, members of the District Commission accompanied by an officer of the Soviet Eighth Garde Armee, Captain Rovensky, paid a visit to Göttern to affirm the distribution of the land of the state domain. In a letter marked "translation," the captain wrote the following:

> Fifty-four hectares were subject to distribution. The tractor and a large inventory were entrusted to the Committee of Mutual Aid. The documents of distribution have been received from the District Commission. They have not as yet been distributed to the farmers. There are now no landless farmers in the town! The owner was jailed. At this moment he is free and living in the town but he has not received anything. Economic Officer of the Soviet Military Commander of Weimar. Hero of the Soviet Union, Captain Rovensky.[22]

Even with land and animals in hand, the "new farmers" were beset with difficulties. It was late in the year for planting. Equipment, fertilizer, seed, and often know-how were in acutely short supply, and even if they had not been, 6 hectares was barely enough on which to survive. Seven hectares was considered the minimal amount that a family of four required. The Göttern local commission had, however, opted to make land available to as many as possible. Some degree of state aid was available. New farmers could receive, on application, 1,500 reichsmarks (RM) as well as credit of up to 4,500 RM for the purposes of building, as decreed by the Soviet Military Government for Germany (SMAD); 6,000 RM was considered the minimal amount required to build a house, an adjoining stall, and a barn.[23] Construction of new properties was out of the question, however, due to a lack of basic building supplies: wood, cement, tiles, nails.

By December 23, 1945, Farmer's Day (*Bauerntag*), the first phase of the land reform had been concluded. In the county of Weimar, 156 private estates of persons designated as large agricultural landowners, war criminals, or Nazi activists had been expropriated and 10,929 hectares of their land divided among 1,312 persons.[24] Much remained to be done. Means had to be found to help the new farmers and the small farmers get on their feet in almost every aspect of agriculture.

Creation of the Farmers' Mutual Aid Association

In early 1946, the first Farmers' Mutual Aid Association (*Vereinigung der gegenseitigen Bauernhilfe*, or VdgB) appeared in the area. Its stated purpose was to help the new farmer and the small farmer resolve their problems by placing at their disposal the resources they required. The first working guidelines, made available on March 21, 1946, instructed potential members as to the whereabouts of machine-lending stations (*Maschinen-Ausleih-Stations*, or MASs), repair shops, seed-cleaning facilities, breeding stations, silos, and mills, all of which had already put their services at the disposition of the VdgB.[25]

One of their first installations was a MAS created in the courtyard of a former estate in Magdala, where expropriated equipment and tractors as well as some Soviet equipment were available for farmers' use. If for no other reason than the fact that large machines required joint use, cooperation between farmers occurred. In time, the Magdala MAS provided services for thirty-one surrounding communities, including plowing, harrowing, rolling, sowing, cutting hay, harvesting wheat, threshing, and digging potatoes and sugar beets.[26]

Having assumed the status of a public entity through government decree on April 5, 1946, VdgB acquired surveillance authority to deter farmers from resorting to black-market activities, redoubling its efforts to make available to the farmer what he needed. Lists of equipment distributed to farmers ranged from harnesses to wooden shoes, twenty-two thousand pairs of which were obtained from the Thuringian shoe industries. The arrival of the shoes in the villages was presented as evidence of factory workers' helping their socialist brothers and sisters in the countryside.[27]

Forging alliances between urban and rural workers was a constant theme in VdgB pronouncements. One point of convergence was the MAS, where most of the technical staff and often the drivers were city workers. Another was May Day festivities organized by local VdgB committees, in which a central theme was the need for unity among all workers, rural and urban. A memo of April 18, 1947, from the Thuringian farmers' secretary to all district farmers' secretaries, titled "May 1st," emphasizes solidarity as well the imperative of participation. It reads, "Farmers shall be members of the May Day Committee to ensure that May 1st is a day of unity for all working people. Flags and flower decorations lift the level of festivity of this day! Farmers, decorate your houses!"[28]

The VdgB opted for not only horizontal ties between peasants and workers but also vertical ties between small and large farmers. The Schorchts had joined the VdgB early on. It was apparent that their membership, and that of others like them, was sought by the VdgB, which, while partisan, recognized a role within it for well-established farmers. For the Schorchts, becoming members meant "going halfway" in the direction of reconciliation. Also, there was the likelihood that, in time, they would all need to make use of the union's services. Under these circumstances, the VdgB would serve as the point of encounter between large and small farmers. Edgar says, "One after another, we all became members. The VdgB developed from the Raiffeisen, which used to be our bank and where we received the fertilizer, loans, and seed, and where we delivered our grain at harvest time."

The Schorchts' advice as breeders was sought by the local VdgB as it began to set up breeding stations. One of them was established on a parcel of the former state domain appropriated by the VdgB. In time, 154 breeding stations were created throughout Thuringia. Like all institutions created by the VdgB and other state entities, the breeding stations were to serve as socialist schools as well as work facilities. A VdgB publication states, "The development and the expansion of the existing breeding stations must be given greater attention by the local committees and the officials of the VdgB, since this must also be seen as an important means of re-educating farmers away from personal egoism towards cooperative behavior."[29]

Some of the antagonism leveled against the "large farmers," so palpable in 1945, had begun to recede. While the situation remained bleak for new farmers, small farmers, and former refugees, they were less apt to see the likes of the Schorchts as culpable. There were new authority figures abroad, some helpful, others meddlesome and coercive.

One whose star continued to rise was Becker. By 1947, Becker had become head of the local VdgB, a member of the District Commission sitting in Weimar, and a member of the Christian Democratic Union (CDU). He was, as Edgar says, always a "mover and a shaker" (*ein Macher*). As a member of the commission, he affixed his signature to a memorandum stating, "In a meeting of the District Commission of the land reform, the Commission states that anybody who causes problems to the new settlers through attacks on the land reform will be expelled from the District and from neighboring towns and cities."[30]

Providing Housing

It was not until 1948 that sufficient building supplies were on hand to allow farmstead construction to get under way. SMAD decree 209, dated 1947, ordered that three thousand new dwellings be created in Thuringia alone. At the same time, it decreed that all former estate buildings be demolished, not only to provide materials for socialist reconstruction but also to destroy reminders of an oppressive past.[31]

The first order of business was to draw up plans for each village, incorporating the new structures. In most instances, the plans were drafted by architects trained in the National Socialist tradition, who had experience designing settlements in the eastern regions. The basic farmstead proposed was the same as the one they had developed for the new German settlers in the late 1930s: a long rectangular building divided equally into living quarters, barn, and stall.[32]

The initial blueprints for Göttern show that seven dwellings were to be erected in a line along one edge of the former state domain, pending the removal of the estate dwelling.[33] An order from the District Commission, enforced by the local commission, stipulated that every farmer in Göttern would be obligated to contribute his labor or cart and horses at some time during the week to the tearing down of the house, the barn, the stables, and the sheds of the former state domain.[34]

Edgar knew every inch of the outbuildings, having played there with Eric Fechte when they were young. The Fechtes leased the estate before the Wagners. That they had formed one of his favorite hideaways was only one reason Edgar did not fancy demolishing the intriguing maze of buildings that made up the estate, but he had no choice. "The whole thing brought about a very bad atmosphere," Edgar says.

First the tiles came down off the barn and the shed and then everything else, except what had been part of the house and the administrator's office above it, which was left standing. That was where the Wagners' Polish prisoners of war (POWs) were housed during the war. Edgar says, "They got nice roof tiles from it, mostly Woellnitzer, made near Jena, the finest roofing tiles made. They ended up all over Göttern. There was Alfons Becker from East Prussia, and he did some stuff [on the sly]. He got the house that was left standing . . . so he didn't need to build. They drew lots to see who would get what, and curiously enough he drew the lot with the house on it. That's how he saw it, just chance. Those were nasty times."

The demolition proceeded a little at a time. Meanwhile, house construction was under way as well, but it moved slowly due to a chronic shortage of materials. On August 25, 1948, the District Commission members in Weimar reported to their Soviet counterparts that although the basements of three of the houses had been dug, construction could not get under way, as "there were no carpenters, no wood, no stones, no glass, and only some roof tiles from the demolition and forty nails apiece."[35]

It was expected that at least some of the 1,500 RM extended to each new farmer by the decree of SMAD would be spent on furniture. The VdgB saw to it that affordable furniture was available. An announcement to that effect said, "In order to provide farmers with furniture, there is [in addition to the socialist new farmers' houses] also socialist new farmers' furniture; it has a country feel to it, all made from solid wood: tables, chairs, armchairs, cupboards, beds. . . . [O]ne kind fits all new farmers."[36] In the end, only three rather than the planned-for seven houses were built in Göttern. Becker had already "won" a dwelling for himself on the old estate. The other three potential homeowners had decided that trying to make a living in Göttern with so little land and poor soil was too difficult, so they left the village. By 1948, the three houses were finished on schedule.[37]

By 1949, with the planned reconstruction finished, the land-reform program was considered complete. In the Soviet zone, a total of 13,689 properties—constituting 3,225,364 hectares, with an average estate size of 235 hectares—had been expropriated. On this surface area, 209,000 new farms, with an average size of 8 hectares, were created.[38]

In the course of the long process that had made the former Dobermann estate unrecognizable, Herr Wagner, his wife, and their five children left Göttern. They settled in Kleinschwabhausen, where Wagner rented a farm and started life anew.

Being a *Grossbauer* and Making the Quota

Although the Schorchts did not always succeed in keeping at a distance from the events of the land reform, their concerns as established farmers were quite different from those of others caught up in it. Fortunate as they were, they considered the years from 1945 to 1949 to be among the hardest they had known. Taxes on larger farms like theirs were raised. Production quotas were established in the form of consignment of crops, "obligatory contributions" to be made to the People's Food Supply.[39] Allocations were based on the number of hectares a farmer possessed.

Those like the Schorchts with more than 20 hectares were categorized as "large landowners," whose quotas were twice as large as those of the small farmers. Edgar says:

> Up to 5 hectares, or even 5 to 10—that was the best quota. Whoever was in this bracket had very little to deliver and could still live quite well. And those with 7 were just in there, and then 10–15 was still possible, and then it got tough—those from 15 to 20 had to deliver twice as much as the ones up to 5, and those over 20 had to hand over even more. We had to give up so much that almost nothing remained for us.
>
> When I think of the potatoes that we had to deliver to Jena, 600 *Zentner* of storage potatoes! We are not potato land here, because we have heavy soil and stones and rocks; harvesting the potatoes is too difficult. We needed the same amount, 600 *Zentner*, to get our pigs fat, and we had to have some for ourselves. How else do you feed ten, twelve people? We also had to deliver 90 *Zentner* of pork. We always had two to three breeding cows, so we regularly had something to sell off, and a little money was always left over. We needed the money to maintain everything. We had the advantage that my father was also a breeder and had good stock, so that the large consignments hurt us less than others with poorer stock or no stock at all. We had bought good stock, high-yield cows that could produce over 3,000 liters of milk—4,000 was the extreme limit—but we had that kind of cow, and others didn't.
>
> Then we also had the apprentices to help. True, we had to start with the kids who didn't know anything, and after two years they would leave us, but it was still advantageous to have the extra help around. We also had day labor from Magdala, women who would help us when there was a lot of work. They had to stop at 4:30 to get home to prepare dinner for their husbands. Their men worked at Zeiss and would be home by 5:30, ready to eat.

Also working in the Schorchts' favor was the right to sell whatever extra produce they raised on the open market. The HO (*Handelsorganisation*) prices were always double the fixed price, especially so for milk. "We used that to our advantage," says Edgar. "Those who were not on their toes the way we were didn't make it through this difficult period. Stoeckel, when he got back from Russian imprisonment, sick, shaved head, wearing wooden shoes, never made it. He had 18 hectares but could only produce the bare minimum. He never got back any butter, and they had four children. That was a bad time."

Pressure from the county, which was in turn beholden to the Soviets, was omnipresent. It came in the form of agents, agitators, and "enlighteners" (*Aufklärer*). Edgar says, "They went out in the field and 'helped.' They were around when they took the [seed] potatoes away from those who hadn't met their quota, and they checked the basements: 'Oh, here you will have to give another 10 *Zentner* on top of your 50, so that the city population will have enough to eat.' That was all right as long as they were fed, but most of it went to Russia. I remember up there in Lössnitz, there was a large potato sorter that was loaded and sent to Russia."

Whoever did not make the quota was not allowed to slaughter. Wally's father, who had 18 hectares, always made his quota. A relative of his with more than 30 hectares was not able to, so Herr Hünniger would kill a pig for them (because he could) and would pass the meat on to them. Edgar says, "No one was supposed to get wind of that. A lot was slaughtered under the table. That sort of thing was harder for city folk. In the first years after the war, so many people came from Jena to work in our fields, not for money but for food. Or they traded their linens for food, because people in the city often had better linens than we in the villages had. My father did not take advantage of this situation. He always said, 'I'll charge the prices we get in peacetime' (*Friedensware für Friedenspreis*). 'Whatever I used to pay for linens, that is how much I'll give you in food prices now.' I always respected him for that."

The Birth of Erhard

The melancholy of the postwar years was happily broken for the Schorchts by the birth of a baby boy, Erhard, on June 24, 1951, Wally and Edgar's first child. The birth took place in the hospital in Jena, where Wally remained with the baby for five days before coming home. Erhard was large at birth, weighing some 9 pounds, 8 ounces, and he breastfed greedily. Within two weeks, Wally was back to her usual farm chores, relying on her mother-in-law to care for Erhard when she was busy in the barn or the field. Elly could not have been more pleased to have a baby in the house again. Even Edwin was reconciled to hearing the baby cry from time to time, as it was his own grandson. Edgar now knew what it meant to be a proud father.

1953: Premonitions

The harvest of 1952 fell far below expectations. Marginal and even well-established farms were pushed to the limit to meet quotas. Unfilled quotas brought threats of dispossession. The Schorchts knew of farmers in the area who had abandoned their land and fled to the West. It took all of Edwin's and Edgar's combined skills to meet their own obligations.

The next year brought even more tension and uncertainty. Uprisings began in Berlin on June 17, 1953, following the shift toward the New Course launched after Joseph Stalin's death. The New Course was intended to slow the 1952 campaign of "building socialism" in the GDR by improving standards of living and lessening political hostility. It failed, however, to revoke the increase in work norms on construction projects, thus leading to strikes and demonstrations.[40] The unrest quickly spread to other cities, including Jena, where Soviet tanks appeared on the streets to put the demonstrations down.

The events of June 1953 had a chilling effect on the Schorchts, casting a sense of gloom on them. They felt increasingly anxious about what lay ahead, wondering whether they were doomed to eventual extinction as independent farmers. In Edwin's case, apprehension expressed itself in his reluctance to make improvements on the farm for the future, something that had always been dear to his heart. Working one day with Edgar after June 17, in the process of rebuilding an old feed stall, Edwin stopped and said, "I don't know what the political climate will be like later on, but it doesn't make sense to finish building." Edgar adds, "We had already caught wind of the cooperatives. We decided not to build anymore."

If the events in Berlin and Jena were not provocation enough, the Schorchts had become aware of plans for the collectivization of the land and believed that sooner or later it would probably occur. When that day came, they hoped to hold out as independent farmers as long as they could.

Commentary

By the time the land-reform program had gotten under way in the early fall of 1945, the immediate anxiety regarding their personal safety that the Schorchts had experienced in the face of American and Russian soldiers gave way to apprehension about the long-term fate of the farmstead. Although the Schorcht property was never in danger of expropriation, the role of the Schorchts as possible *Grossbauern* (capitalists, exploiters) was a made-to-order target for Socialist Unity Party of Germany (SED) propaganda. The attacks came in waves, depending on the SED's political needs. In 1948, the incriminating accusations against the likes of the Schorchts were heard again, identifying for the small farmer the source of their frustrations with the land-reform project even as the program was coming to an end. It was apparent that the political gains to be achieved from the program were as important, if not more so, than economic and social ones. The politicization of the village made this clear.

With the creation of the local land-reform commissions, empowered with executive as well as deliberative powers, a kind of radical democratization was imposed. The commissions were assigned a daunting task—not only the identification of property eligible for expropriation but the physical seizure of it as well. Little wonder that there was equivocation within the Göttern Commission in regard to seizing the Dobermann estate. Never before in Göttern's history had persons with such meager economic interests been called on to make decisions of such consequence in the political arena. We may speculate about the range of doubts members of the Göttern Land Reform Commission might have had: from reluctance to act against a man of Wagner's status, to suspicions concerning the Communists' crusade for land reform, to anxiety about their own ability to become independent farmers. The pro-Wagner faction, among whom were well-established farmers like the Schorchts, who found no justice in the actions of the commission, was not sufficient to deter expropriation—little wonder, considering that behind the local commission stood the county commission in Weimar, which in turn had the support of the Soviet Military Administration of Thuringia. It was important to the KPD that the commission's decision should appear to be the action of an aroused peasantry, thus serving to make its members complicit in an act of class warfare.

The Wagner case is worthy of consideration, for it represents in microcosm the articulation of subjectively experienced material needs (the demand for property among the evacuees) alongside the political demand (at the state level) for the ideological mobilization of the citizenry. Apart from the intrinsic merits of the case, the decision to seize the state domain reflects the Wagners' victimization. Their departure left a vacuum in the social matrix of Göttern into which one of the larger peasant families might have moved if circumstances had been otherwise. But in the fall of 1945, the traditional criteria of preeminence carried little weight. When the Wagners left, what the Schorchts lost was neighborliness, as they had experienced it before the war.

If the intensification of the political process and its subsequent realignment numbered the squire of Göttern among its victims, it counted among its new leaders a refugee from the East, Becker. Becker, quick-witted and vocal, was able to capitalize on the dissension within and outside the local commission. In the election of November 6, he was catapulted to its chairmanship, after having served as the nonvoting secretary. His "dark forces at work" letter is a masterpiece of innuendo and vengefulness. At its closing, he unfurls the banner of the protector of "the poor and the poorest" against all those who "want to hinder the good work of the *Bodenreform*." Although the Schorchts are not mentioned by name, Edgar was among those who "brought out another two wagonloads . . . without saying a word to me."

In the few months since his arrival in Göttern, Becker had become the village's moral entrepreneur, casting his net of aspersions over those he held wanting. Such were the changes in the political climate of Göttern. Becker continued to climb, becoming the chairman of the local VdgB and a member of the District Land Reform Commission. The last we hear of him is Edgar Schorcht's thinly veiled reference to the manner in which Becker "won" ownership of the remains of the Dobermann mansion. Becker demonstrated a truth residing in the generalization that in periods of uncertainty, men of ambition and innate talent can win positions of influence. At the height of his powers, Becker effectively bridged the space between local and district spheres of power, at a time when core and periphery were, in turn, being emphasized.

If the land-reform program served as a viable mechanism for mobilizing the rural proletariat, its effect on farm production was markedly adverse. Figures show a decline in crop yields in the Soviet zone to a fraction of prewar levels. War devastation was to blame, as was the Soviet removal of equipment and livestock, but so was land reform. Fall planting was disrupted by the seizure of property. Capable estate owners like the Wagners and their overseers were banished. Disruption was created by inexperienced land-reform commissions. Small farms proved to be economically inefficient, and the "new farmers," especially former city dwellers, were largely incompetent.[41]

Regardless of whether problems arising from small farm sizes and inexperienced farmers were foreseen, the VdgB was created to resolve them. All the resources a farmer required were accessible in the machine-lending stations and other services made available by the VdgB. Nevertheless, there were those who found the challenge overwhelming due to age, health, inexperience, economic difficulties, and lack of adequate shelter. The return rates for land in Thuringia and in the Weimar district were particularly high.[42]

For new and small farmers who remained to become independent cultivators, however, the VdgB services fostered mutuality and dependence. The large machinery—combines and threshers—were designated for joint use.[43] Dependency on each other and on the VdgB not only was an immediate solution to problems faced by the new farmers but also was designed to serve as ideological preparation for the step toward collectivization projected for the 1950s. Participation in the services offered by the VdgB created a contradiction between the aspirations of independent farming on one hand and dependency on the other, a contradiction to be resolved by enforced, institutionalized cooperation, a future the Schorchts had already foreseen.

10

Irmgard Defects to the West

Uncle Walter's Invitation

RMGARD HAD NEVER consciously thought she would leave Göttern permanently, even though her life there had not worked out the way she had hoped. She was devoted to her parents, acutely aware, as she was, how hard they worked for the good of family and farm. Yet she could not help but remember the occasions on which they had thwarted her plans to achieve what she had wanted for herself. At the time, she knew that they had done what they believed was right. Then when Edgar married, what had been obvious all along became all the clearer: there was no place for her at home. Irmgard says, "My problem was that I wasn't egotistical enough. I should have said [much earlier], 'So that's it. I'm doing what I want to do from now on.' As for men, there were lots of men I could have married, but my expectations were never low. My mother said that I had one for every finger, I could have my pick, but I didn't want any of them. The ones I liked, my mother didn't. One must realize, too, that men my age, and many dear acquaintances, had died; it was the war generation. Then my uncle Walter wrote and said that I should come to Pirmasens."

Walter Michel's Travails

It was the winter of 1954 when Walter Michel wrote from Pirmasens—in Rheinland-Pfalz, near Saarbrücken—where he had gone to live following his denazification (*Entnazifizierung*). Irmgard had once visited her uncle for eight weeks during the war, when circumstances were quite different for him. He was then living in Zell an der Mosel, having been sent there from Weimar to be the local finance director and concurrently serve as the Nazi Party's local director. He had become a person of some importance and means, and Irmgard was proud to be related to a man of such consequence. He was married to a woman thirteen years his junior, with whom he had four children. With the end of the war, his fortunes changed drastically. Arrested as a Nazi Party official, he was incarcerated, underwent a trial at Ludwigsburg, and then was subjected to denazification after having been transferred to a camp near Stuttgart. There, he enlisted as a construction worker, learned to be a mason, and in time trained to be a foreman.[1]

Believing that Uncle Walter had suffered at the hands of destiny much the way she had, Irmgard went to visit him again in 1948 with her family's blessing. She brought with her, illegally, a suit belonging to her grandfather Louis so that Walter would have something to wear. At the time, Walter's wife was living nearby. She wanted a divorce, claiming that she could not remain married to a mason. Irmgard says, "He was half-starved, but that didn't keep his wife from going shopping with his wages. She said, 'Look at the lovely leather jackets.' She wasn't thinking about his welfare. She knew how he loved their children, and she blackmailed him with them. Later, she moved away and took up living with a veterinarian who was married. She split them up, too," Irmgard recounts.

When Walter was eligible for release from detention, "Many people came from Zell to speak in his favor," says Irmgard proudly. In the end, he was released without any further obligation. "When he began looking around for work, he was offered a job as a tax adviser for the state. He asked his sister Elly for her advice. She was the closest relative he had left. She told him, 'You have four children, and you are no longer so young. You had better go work for the state, where you'll have a secure position.' Her brother took her advice and was then stationed in Pirmasens. Irmgard says, "Pirmasens? We had to look it up on the map. But then we saw that it was near Kaiserslautern, and we knew about Kaiserslautern from Pfaff sewing machines. Meanwhile, he and his wife had obtained a divorce, and he had met a war widow in Stuttgart, and they married. She had one daughter, and he had custody of two of his children."

Irmgard Arrives in Pirmasens

When Irmgard stepped off the train at the Pirmasens station, she was struck by the changes that had occurred since her last visit in 1948. Pirmasens, an industrial city famous for its shoe factories, had been severely damaged by Allied air raids. Even then, three years after the war, there was evidence of devastation everywhere. But now in 1954, Pirmasens was on its way to being transformed. There were signs of construction and reconstruction at every turn, accompanied by a feverish spirit of activity and optimism. Irmgard found it intoxicating. It was a relief to be away from the constant smell of animals and the buzz of flies around the manure pile in the courtyard at home. She says, "I had still left the door open for going back home, but then I thought, 'No, I won't go back home!' It wasn't easy allowing myself to decide that, and I never wrote my parents about that either."

When Irmgard accepted her uncle's invitation to join him in Pirmasens, she had no place to stay. At first she slept on a mattress in her uncle's living room. Then he found a job for her as a secretary in a factory that made synthetics in nearby Heltersberg.[2] Sitting at her typewriter near the offices of the director and the other managers of the thriving firm, she used to think, "Oh, if someone from home were to see me now! But I needn't have been ashamed [working for the capitalists]."[3]

After the first six months, Uncle Walter was transferred to Bad Kreuznach, where Edgar had been incarcerated as an American prisoner of war (POW). There, he built himself a house, retired, and ultimately died. Before Uncle Walter's death, Edwin paid him a visit. By then, the political differences that had separated them did not seem important anymore. Erhard remembers going with his father and grandfather on that occasion. He thinks it must have been 1957 because of what Uncle Walter said when he asked him what he used the gun in the corner of the room for: "I wanted to use it to shoot down Sputnik," he had answered.

Irmgard was now alone in the West. She moved into the neighbor's spare room, where she took her meals. It was a good distance from the factory, especially to and fro, four times a day. She did not have the money for bus fare but found that she could make her way there and back on foot almost as quickly; besides, she met more people that way. She left the house at 6:00 in the morning to start work at 7:00. Work finished at 5:00 except when her employer kept the employees late, and there was a half day on Saturdays. Irmgard says, "Everyone worked hard. They say the *Pirmasenser* are rough, but rough with feeling. They were always ready to help me. It just depends on how you deal with people. But I never depicted myself as a refugee. They were just good people. When I worked in the factory, there was real human warmth; people were nice to one another."

Irmgard was earning 180 deutsche marks (DM) a month at first. She found it hard to put aside anything but, nevertheless, began to send home packages of sweets for everyone as well as hose for her mother. It gave Irmgard pleasure to do this. She was beginning to believe that, at last, things were working out for her.

The Birth of Norbert

Wally and Edgar's second child, Norbert, was born on April 2, 1955, in the hospital in Jena. After a recovery period of four days in the hospital, Wally joyfully returned home with her newborn. Edgar was glad to have another son to be at his side in the years ahead. Edwin felt doubly reassured about the future of the Schorcht farm now with two grandsons—provided that the Communists would leave them alone. Elly was delighted to have another baby in the house. Her dreams of grandmotherhood with Edwin at her side were coming true. Erhard, now four, had little or nothing to do with his baby brother, evincing more disdain than pleasure at his sight.

The Ring

It was about the year that Norbert was born that a family ring was lost, later to be found. Edgar, who told the story more than once, relates it like this: "My mother had a wedding ring. She was wearing it even though it was weakened from work. My father didn't have one, and I haven't had one for a long time, because they wore out. One day in the cow stall, my mother's ring slipped off her finger. She looked and looked through the straw where she had been working but couldn't find it. Upset, she finally gave up the search. Two years later, the ring was found. We were up in the field picking potatoes. My wife saw one of the hired girls reach for something on the ground and put it in her handkerchief. The girl didn't say anything. My wife asked what she had there. She said, 'A ring. I found a ring.' My wife looked at it and saw the inscription, the engagement date of my mother—not only the date but also the name, 'E Sch. 1919.' My wife took the ring and gave the girl some money. When my mother saw the ring, she was very happy. Later she gave it to my sister in Pirmasens, even though my wife had found it. My sister had some precious stones set in it and wore it. Several years later, when my sister came in from Pirmasens for her annual visit, the story of the ring was told one evening. Then my sister came to know that it was my wife who had found the ring. My sister said, 'I have worn it for all these years, now it is your turn to wear it.'"

The story is told by Edgar as if it conveys a truth: material things in life, like rings, are transitory; they wear out, they are lost, but relationships endure.

Edgar Assumes Control of the Farm

By 1958, it seemed as if the Schorchts had put the most difficult of the postwar years behind them. Edgar was gaining ever more confidence in his ability to match his father's standards in farm management, at the same time aware that in some areas of cultivation techniques, especially fertilization, he was more knowledgeable than his father. The apprentice program was running at high gear; consequently, labor shortage was never a serious problem. Even so, they missed Irmgard for her ability to round up women in the area during harvests, but even more for her organizational skills and her aptitude for keeping the books. Irmgard's mother referred to her as their "foreign and interior minister," taking care of their "bureaucratic needs."

Irmgard remembers occasions when she fronted for her father in dealing with Communist officials. She says, "I remember after the war, when you had to get permission to slaughter an animal. You had to go to Weimar to get that. My father couldn't stand those people, and he let them know it. If he knew them before the war, he would say that they were playing themselves up to be big fish now. I told him, 'Father, you stay at home; leave the boys to me.' I planned out my strategy. . . . They are freshest on Mondays; let them have read their newspapers before we start. Shortly after 9:00, I sauntered in; a half hour later, I left with the permission slip, and my father was so mad. 'How do you do that?' I'd tell him, 'Father, when you go in there with a sour puss. . . .' It wasn't pleasant, but you had to do such things, and I did it much faster than he could. I always used to say, 'No need to bother an angry dog.' But I always got what I needed."

Now that Irmgard was in Pirmasens, Edgar assumed the role of protecting Edwin from the local Communist authorities, although he found it hard sometimes to keep his temper. It seemed as if middle farmers like the Schorchts were always under attack. Nevertheless, each year they increased their crop and milk production, with consequent income gains but likewise appreciation in their tax payments. Edwin and Edgar consulted among themselves how best to incur a lesser tax burden. They finally decided that Edwin would lease the farm to Edgar, thus diminishing Edwin's taxes. After finding a notary for whom this transaction did not represent a transgression of the law, the deal was closed. Edgar had become responsible for the farm's management even though its ownership still resided in Edwin's hands. Edgar assumed his new role as if it were to be his lot for the foreseable future. He had reason, though, to doubt that to be the case: collectives were being formed throughout the district of Weimar.

The Birth of Roswitha

On August 13, 1959, Wally gave birth to her third child, a baby girl, Roswitha, born in Sophienhaus in Weimar. On August 11, Wally was in the fields helping with the harvest, and on the day before she went into the hospital, she joined in the end-of-the-harvest party. Wally could not have been more pleased to have had a girl. As much as she loved her two boys, she wanted to raise someone who would grow up to be like her, someone with whom she could share confidences in the years to come. Edgar already had some inkling of the blessings that a father-daughter relationship could bring. Elly, for her part, believed that with the birth of Roswitha, her life had been made complete; she now had a granddaughter to watch grow into young womanhood, if she was fortunate. Edwin reveling, albeit reservedly, in being called Opa by his grandsons now looked forward to the time when Roswitha's small voice would join the others. Erhard stood his distance yet permitted himself more tolerance toward his baby sister than he had toward his brother. Norbert almost from the beginning revealed an affinity for his little sister and she for him. Heartfelt greetings were forthcoming from Irmgard in Pirmasens.

Irmgard Marries

In 1960, Irmgard married Karl Stüber in Pirmasens. Irmgard and Karl had met through a mutual friend. Karl, born in 1911, the youngest of four, was twelve years Irmgard's senior. Karl's father, who had owned a store and considerable land in Pirmasens, died in 1917 when his horses flipped over on him while delivering his wares. The first day Karl went to school, he went alone. His mother was in mourning. She could not manage the store by herself, so as Karl grew up, he worked at his mother's side. It was Karl who drove the car to make deliveries. He was one of the first of his generation in Pirmasens to have an automobile.

When the war came, Karl served on the Russian Front. When he was reported missing in action, his mother, convinced that she would not be alive when he returned, asked her sister-in-law to look after her son when he came home to Pirmasens. Karl wrote, but his letters from Russia never arrived. Irmgard says:

> The others all cried, but my sister-in-law said, "I feel it—he is still alive, and he will return to us." And that is what really happened. She had a sixth sense.
>
> She told me that on the day of the bombing raids on Pirmasens, her husband was visiting her mother. She had called him and told him to bring everybody back to their place. He said not to worry, but he had no intention of doing so. She yelled at him for not being a responsible member of the family. Angry, he hung up, got in his car, and drove home . . . but alone. A short while later, the house was bombed. Karl's mother was killed. When they found her body—she died from one of those bombs that make your lungs burst, I forget

what they are called—they found her pocketbook and the ring that she had told her daughter was to go to Karl. She carried it with her everywhere she went. When Karl finally returned, he got the ring— that was the last memory he had of his mother.

One evening Karl was driving Irmgard home, several months after they had met. As if out of nowhere, a car approached with its high beams. There was a head-on collision, and Irmgard's head hit the dashboard. The skin of her forehead peeled back, and she had to remain in the hospital for several weeks. Every day Karl came to see her, bearing flowers and other gifts. Irmgard had never been courted before in this manner. By the time she was discharged, she and Karl were engaged to be married.

Irmgard talks now, several months after Karl's death in August 1992, of the affection and love she felt toward him. She says, "He was not only industrious but personable and reliable in every way. He always made sure to take care of his appearance. He had an aesthetic sense; I particularly loved his beautiful hands. He could be very serious at times and then he would be quiet and not speak for a long while. I was always very impressed with how many things he was capable of doing. He was technically gifted. It was very important for him that things be right. One of the first things I gave him after we were married was a level. Everything had to be level. We would go on vacation, and if the light-switch plate was not screwed on properly, he would fix it."

It was a small summer wedding. Karl's sister and husband were in attendance, as were Irmgard's father and mother. Edwin and Elly were delighted that their daughter had finally found the right man. For years, they had worried about her, wondering whether she would ever find him. They liked Karl, even though they could not understand his dialect. He was a large person, whose reassuring presence promised that he would look after their daughter. Edwin and Elly believed that their lives were more complete. It was the last time Elly was to visit Pirmasens.

Commentary

Irmgard's move to the West, as an unmarried younger woman, was an act of individuation that conflicted with her mother's sense of decorum. The fact that Irmgard had gone to Pirmasens under the tutelage of Elly's brother, Walter Michel, undoubtedly made the relocation more tolerable, as had the harsh realities of recent history. Elly had most likely become more tolerant of behaviors that before the displacements of the war would have been unacceptable, yet the disparity in expectations that separated mother from daughter were slow in dissolving.

With her upbringing in a small rural village, Elly's horizons were restricted from the very beginning. Born on July 12, 1893, a child of Wilhelmine Germany, Elly was twenty-five when the kaiser abdicated in 1918. She came of age in a time of pronounced social propriety, when an essentialist perception of gender differentiation was the norm, along with an affirmation of the centrality of home, mother, and children, especially among the social strata to which Elly's parents aspired. She attended the local one-room elementary school. From what Edgar and Irmgard have said about their mother's industriousness and quick-wittedness, she was a good student. She completed the required eight years of school in 1907, when she was fourteen. There, she formed a respect for learning and the arts, especially singing. She was undoubtedly influenced by the emphasis placed on order, discipline, and obedience to male authority, traits that stood in some precarious balance to her outgoing nature.[4]

When Elly was sixteen or seventeen, she spent a year in a cooking school in Jena, where she was exposed to the requirements of feminine etiquette as well as to the niceties of more refined cooking—behavior and knowledge designed to enhance a young woman's marriageability. For the next thirteen years, Elly helped her parents in the Gaststätte, "Zur Linde," waiting for the proper man to ask her to marry him.

A picture of an eighteen-year-old Elly from 1912 shows her in a long white dress. The choice of white had recently become acceptable for a young woman of the countryside, white being previously reserved as the color of preference for the aristocracy and then for the bourgeoisie.[5] The picture suggests a young woman ready to embark on a number of different lives. In truth, only one was accessible to Elly: the need to demonstrate a preference for a man to marry among those available in Göttern and its immediate surroundings.

In the photograph taken on her wedding day, Elly is dressed in black, the customary wedding-gown color for a person in her station. Elly's acceptance of Edwin's proposal reaffirmed for her, the family, and the community a commitment to a life of domestic dedication.

The Weimar Republic was three years old at the time of Irmgard's birth. The outward appurtenances of Imperial Germany had been swept away, leaving intact much of the institutional structure as well as the traditional exclusionary and deferential patterns of everyday life. Existence in small villages like Göttern went on much as it had before. Nevertheless, some changes in social customs did make themselves felt in the countryside as well as in the city. New standards of fashion allowed women more freedom of movement by shortening the skirt, loosening the waistline, and lowering the neckline, an aesthetic in keeping with the image of the "modern woman." A formal picture of the Schorcht family taken circa 1923 shows Elly wearing a striped blouse open at the neck, in contrast to the high-collared dress she wore in 1907.

The first elections for which Elly was registered in Göttern were the municipal and county elections of 1922.[6] Women's suffrage had been granted in the Weimar constitution three years earlier. Although we do not know how she voted in the National Reichstag elections of September 14, 1930, we may conjecture that she joined with ninety-eight others from Göttern, as well as with her husband, in casting her ballot for the Christian-National Peasants' and Farmers' Party. Likewise, we may surmise that in 1932, Elly voted for Paul von Hindenburg, who was running against Adolf Hitler and the communist Ernst Thälmann, in the second round run-off of the presidential election. It is unlikely that Elly voted in either of these two elections with the progressive needs of women in mind, instead using her ballot to retain the status quo.

When the Nazis came to power on January 30, 1933, there was little reason for the inhabitants of Göttern to believe that a telling moment in German history had come to pass. Life continued much as usual. A little more than a year later, on February 10, 1934, Elly Schorcht joined the National Socialist Women's Organization (*NS-Frauenschaft*, or NSF), an adjunct of the Nazi Party, seven years before her husband became a party member. The exclusionary preamble of the NSF to which she affixed her signature echoed the sentiment of the day. It reads, "I hereby declare my admission into the National Socialist Women's Organization. I am of Aryan-German descent and free from any Jewish or colored racial elements, belonging neither to the Free Masons or any other secret organization, and promise not to become one during the duration of my membership in the National Socialist Women's Organization."[7] She may have perceived the NSF as the natural successor to the spinning bee (*Spinnstube*), while hoping for advantages other than those offered by the traditional gathering. As Gerhard Wilke says of the NSF and the League of German Girls (*Bund deutscher Mädel*), they "had not only an ideological retraining function. They also allowed women to travel beyond the narrow confines of the village, brought them into contact with women from other regions, and made them cross class boundaries."[8] Yet beyond touching on such experiences, the NSF was not a movement for the liberation of women, for it preached the virtues of hearth, home, and motherhood. It may be that Elly and other women like her saw in the NSF an unambiguous supporter of traditional values—that is, Elly may have been as ambivalent in regard to the virtues of modernity as were the times in which she lived.

In 1934, the year that Elly joined the NSF, Irmgard was eleven years old. She was raised as a farmer's daughter, helping in the kitchen and the house and working with her mother to turn out a seemingly endless supply of cake and bread. She learned the ways of the stable and field as well, becoming an adept milker. She was even allowed, on occasion, to lead her father's prize bull, Kronprinz, around the *Hof*. According to Irmgard, however, farm life held less attraction for her than it did for other girls in the village. She liked going to school, where she did well. Whether it was her teacher, Walter Venus, who inspired her to excellence or whether she would have been a high achiever anywhere is difficult to say. It was, most likely, character and circumstances that combined to make her a good student.

In the ensuing years, Irmgard had two experiences that may have stimulated her imagination. The first was the special attention Herr Venus paid her as a favored student, teaching her how to write, from dictation, the letters he sent as a Nazi Party official. Some of his correspondence was to persons at the highest level in the party. It appears to have been flattering to Irmgard to have been chosen for such a task. It may have also given her a glimpse of a world beyond Göttern.

The second informing experience occurred while attending the School for Home Economics in Jena when she was fourteen. As fortune would have it, she found herself a room in a villa that exceeded her expectations. The fin-de-siècle splendor of the mansion, the cosmopolitan boarders, and the spacious garden opened her eyes to a life she had only a glimpse of before, on passing through Jena. She was aware that not all women were destined to be farmers' wives, yet she knew very few who were not.

At the end of the school year, she expressed her desire to stay on and to seek admission to the *Gymnasium*, knowing that her parents wanted her at home on the farm. In the face of their opposition, she had to renounce once again her dream of further academic study. At the time, Irmgard must have acutely felt the conflict between wanting to be a good daughter and wishing to realize herself as a person. It is doubtful that Elly experienced such cross-purposes to the same degree, if at all, either because of the characterological differences between mother and daughter or because of a generational shift in social expectations.

Irmgard became a National Socialist Party member on September 1, 1941. According to Irmgard, it happened automatically when she reached eighteen. Irmgard was the first person in the immediate family to join, although Walter had joined in 1931. The party welcomed her, but Irmgard would learn in time that what had seemingly served her uncle so well was not the answer for her. The NSDAP was unrelenting in its opposition to women's assuming leadership roles at either the local or the national level. The party was just as obdurate in insisting that a woman's duty lay in providing comfort to a man in marriage and succor to a child in motherhood. It was clear that Irmgard ran the risk of becoming a pariah in her own land if she did not accede to the party's expectations.

During and after the war, Irmgard was constrained by the realization that because of her sex and the circumstances of her life (as a farmer's daughter), few options existed for her in the East. It would be the forbidden Germany of the West that would offer the answer to Irmgard's dilemma, and Uncle Walter provided the catalyst. Irmgard was thirty-one when she received the letter from her uncle asking her to pay him a visit. By then, there was no more talk of Irmgard settling down and marrying in Göttern. Unable to work on the farm and with no room for her in the house that was to belong to Edgar and Wally, Irmgard made the decision to leave. By then, she had learned to recognize what had become a lifelong pattern of behavior: the renunciation of what she wanted in deference to the wishes of others. She says, "I know that now. . . . I should have looked out for my own interests. I let that happen." Subsequently, she is able to say, "So that's it; I am doing what I want to do from now on." And she never turned back.

Departing the GDR for West Germany was an act fraught with political implications even when the reasons for going were personal, as they were in Irmgard's case. Although the invitation from Uncle Walter was the pretext for leaving Göttern, what had been described as a visit was, in fact, an escape once Irmgard settled in Pirmasens. Irmgard's action multiplied a thousandfold became the flood that Ulbricht intended to staunch by the building of the Berlin Wall in 1961. After 1961, contact between Irmgard and her family in Göttern was restricted to censored correspondence, packages Irmgard sent, Edwin's annual trip to Pirmasens following his retirement, and journeys Irmgard made alone or with Karl to Göttern when permitted. In addition, Irmgard's residence in the West was not without its repercussions for the Schorchts of Göttern. Having a "West-Kontakt" in Pirmasens was a political fact useful to the local branch of the Socialist Unity Party of Germany (SED) when it wished to point a finger of suspicion in Edgar's direction.

Nevertheless, history served Irmgard well. Settling in Pirmasens with a job as a secretary in one of the large plants provided proof of her ability to live free from family and farm. She sent packages back to Göttern filled with things to wear or to eat. Irmgard came to feel more in charge than she ever had before.

In the commercial and industrial city of Pirmasens, Irmgard was able to find a solution to the dilemma that marriage posed for her in Göttern by identifying a suitable man to marry without paying the price of becoming a farmer's wife.

In her defection to the West, Irmgard succeeded in minimizing the long-term sense of personal insufficiency imposed by gender and inheritance by exchanging a marginalized East Germany for an ascending West Germany to which she could dedicate her allegiance and which she could claim as her new home—a vantage point from which she could be less a child and more an equal to her mother and father.

11

Collectivization of the Land, LPG I

From Land Reform to Collectivization

THE DECISION AT WAR'S END to subdivide East German estates into small farm holdings rather than collectivize them reflected the political belief that the necessary support of the German agrarian workers, the marginal farmers, and the refugees could be won only by providing them with the means to become independent farmers, not by organizing them as workers on large industrial estates. It also stemmed from the fact that neither the material infrastructure nor the trained cadres were yet in place to carry out a program of agricultural collectivization. The creation of a class of small farmers was seen as the expedient solution, provided the peasant could be convinced of the eventual necessity and inevitability of collective farming. Only by collectivization, it was held, would the contradiction between the privatized small agrarian producer and the socialized industrial worker be resolved.[1]

The First Collectives in the County of Weimar

With the completion of the land-reform program in 1949, plans for the formation of collective farms were already well advanced by the early 1950s. The bad harvest of 1952, the inability of thousands of farmers to fulfill their quotas, threats of dispossession, the exodus to the West, and the amount of still-untilled land persuaded the authorities to quicken the pace of implementation. The Socialist Unity Party of Germany (*Sozialistische Einheitspartei*, or SED), formed in 1946 by the unification of the Communist Party and the Social Democratic Party, began announcing the organization of agricultural workers and new farmers into agricultural production cooperatives (*Landwirtschaftliche Produktionsgenossenschaften*, or LPGs) a year before the uprising of June 17, 1953.[2] On July 13, 1952, the first agricultural collective in the county of Weimar, the Ernst Thälmann LPG, was founded in Isseroda, some 20 kilometers west of Göttern.[3] Named for the founder of the German Communist Party, who was killed in Buchenwald in 1944, the Thälmann cooperative was a type I LPG, meaning that only farmers' land was worked communally, leaving livestock and equipment under private control. In September came the establishment of three additional LPGs, including one in Magdala. By the end of 1952, there were eight LPGs in Weimar County, involving 83 farms, 177 members, and 624 hectares. This first step in the grand design to eventually industrialize the countryside involved a small fraction of cultivatable land.[4]

By December 1953, the statutes for LPG types II and III were decided on. Type II provided for the collective use of arable and pasture land as well as machinery and tools, while type III was characterized by the collective use of all means of production, including land, machinery, livestock, and buildings. The LPG III, the "perfect type," came to be looked on with official favor. Yet even in an LPG III, a minimum of private property was preserved in the form of a half hectare of land for growing fruit and vegetables to be sold for profit on the open market. An additional enticement to induce membership in the full collective was the permission to retain two cows or horses, two sows, and as many sheep, goats, and poultry as a farmer could support on this half hectare of privately cultivated land.[5]

In theory, joining an LPG was voluntary, but pressure by the SED was, in fact, necessary to build membership. The response of farmers all through East Germany, not the least in Thuringia and in Weimar County, was half-hearted at best, despite the herculean propaganda efforts to win them over. At least nine thousand East German farmers fled to the West in the first months of 1953. The unsuccessful uprising in Berlin on June 1953, while primarily an expression of the urban working class, was also undoubtedly reinforced by the peasants' resistance to collectivization.[6]

Following the sobering events of 1953, the policy of the government was to go slow in promulgating LPG membership. By the end of 1959, three-quarters of useful land in the Kreis Weimar was still in private hands.[7] In comparison with other members of the Warsaw bloc, only Poland was lagging behind East Germany in regard to the socialization of its agriculture.[8]

The "Socialist Spring" of 1960

By 1960, however, whatever restraint the government had theretofore practiced was cast aside in favor of a "crash program" to achieve 100 percent collectivization. Six years had passed since the uprisings of 1953, with the regime now feeling sufficiently secure to apply greater coercion if necessary to achieve its longstanding goal. As of 1960, there was also in place a sufficient degree of technological infrastructure to enable the creation of farms of immense size, in regard to plant and animal production, with the hopes that a rational use of the most modern mechanization would maximize food production. Not the least important consideration was the ideological conviction held by the SED, following Soviet theory and practice, to bring about a major social transformation in age-old modes of human behavior and relationships—a veritable moving of historical mountains.[9] The decision to speed up collectivization by all means possible was made at the Seventh Session of the Central Committee of the SED, held on December 10–13, 1959.[10]

In the first four months of 1960, a total mobilization of SED cadres occurred to make certain that by May 1, the date of the "Socialist Spring," there would not a be a single independent farmer to be found in East Germany. Coercion, although short of physical force, was intense and thorough. Reluctant farmers were threatened with prosecution for illegalities or told that those left remaining outside the productive cooperatives would be denied access to seeds, artificial fertilizers, or parts and repairs for farm machinery. By April 14, well ahead of the deadline, the Karl-Marx-Stadt district, the last to report, announced full compliance.[11]

Göttern's First Collective

The Schorchts were determined to remain a private farm for as long as they could. In the three years that he had been in charge of the farm, Edgar was convinced that he could maintain his momentum of increased production, despite the roadblocks the authorities kept throwing in his path. As far as the Schorchts were concerned, collectivization was well suited to the needs of the small farmer, who benefited from the pooling of resources it provided. For families like theirs, which had been farming for generations, collectivization would destroy an enterprise proven to have been productive. Edgar says, "We were doing so well, we didn't need a cooperative." Nonetheless, the Schorchts found it convenient to use a large tractor from the machine-lending station (MAS) to turn their heaviest soil.

In 1958, the first collective in Göttern was established. It was for new and small farmers, some of whom had received land during the land reform. Named LPG *Gemeinschaft* (community), it was founded by Gerhard Becker, a relative of Alfons Becker, and included among others Gerhard Rothe, the former mayor. Four years later, seven other small farmers joined.[12] Because it was a type II collective, land and equipment were used communally, thus maximizing the benefits of pooling scarce equipment. Livestock remained under private control. Like the Schorchts, however, large farmers in Göttern continued to operate independently. Yet as the spring of 1960 approached, the inevitability of collectivization for them was clear. The transition was made a little easier for the Schorchts by the fact that in 1958, Erich Hünniger— Wally's father, whom Edgar respected—had joined a collective in Bucha.

Farmers in Bucha were more amenable to collectivization than those in Göttern. As Edgar explains it, "My father-in-law in Bucha was alone. My wife was here, and her sister Gertrude also was married and gone, and my brother-in-law was missing in action and he didn't come back, even though we were all sure that he would appear again. Before my father-in-law joined the LPG in 1958, I used to take the tractor out to Bucha to help him out. I mowed and helped drive the feed in, harvesting in the fall. He had lost his tractor to the SS [*Schutzstaffel*] during the war. Just before the Americans came, the SS took his nice Normac, a small diesel tractor, away. He was left with two horses, and they took one of them, too. In 1958, he looked around at what other people were doing, and after a while said, 'Why should I work alone anymore?' and so joined up."

Referring to his own farm, Edgar says, "Even so, we could—and should— have held out longer; it wasn't of our own accord that we joined up. Our farm was doing so well, we didn't need to cooperatize. But a lot of people had been ruined by politics and the delivery quota system, and they had to cooperatize. The pressure was there. [The agitators] worked people over until they finally gave in. We were virtually gagged—not enough fertilizer, not enough seeding material—until we were finally 'prepared' to enter. It got to the point where the political pressure was so strong that we said, 'There's no point in resisting anymore.'"

Farmers who took the initiative to organize themselves received support from the state. The Schorchts were aware of the situation in nearby Schorba, the first totally cooperatized village in the area. Edgar says, "The 'enlightener' (*Aufklärer*) had done a good job there, and the farmers of the village all signed up—all big farmers, too—and for that they got their cultural center right in town, a big ballroom, a *Gäststatte* kitchen, everything. They had all sorts of advantages. Where the farmers agreed and the best farmer became the chairman, things worked out. That's how it was in Schorba. But we were too late to get all the money they got."

The Schorchts and the LPG Zur Linde

Collectivization among larger farmers began in Göttern when they started holding meetings to discuss ways to organize themselves into an LPG I. Edgar says, "Even though we were not politically active, we said we [the farmers] should try to take over the leadership of the cooperative. The political side of it was decided in Weimar anyway, so leave the farming up to the farmers."

The general consensus was to come up with the least-constricting organization possible. Edgar says, "In March 1960, we started having meetings—some here in my kitchen, then over in someone else's kitchen. The way it was decided was this: we would group three to four around each of the eight farmers who had tractors so that everyone would end up on a pretty even footing.

"I had Schwarz and Becker with me, and my cousin—nobody else wanted him; he was somewhat negligent, so we had to do it together. Everybody was still in charge of his livestock; everyone would do his fodder in the morning on his field. But in regard to crops, we had to meet together, work together, and bring in the harvest together. I was able to organize it in a way as to not discriminate against anybody. We finished the grain crops within one day of each other, then we had to do the potatoes, also together. There was one of the eight of us who had a good farm, 30 hectares, but he was a little peculiar, and they couldn't get it in all in one day, as there were only two of them. So we all had to help out together, but those were his potatoes—he had the right to sell them off. It was only as of October that the cooperative really got under way; that is when we started putting the fields together."

Edgar continues, "It wasn't that simple, because the fields—we called them 'towel fields' (*Handtuch-Flächen*)—had to be turned into big fields. We had forty individual farms and had to make a few out of so many! And then had to get the fields all in sync because of crop rotation. . . . Some had rotated crops, and some hadn't. The first summer, barley can be planted anywhere. Potatoes can be planted anywhere, although they don't do very well here; wheat is a little more particular, and so are turnips. It took a long time before the fields began producing similar results. It took about five years to [get to that point]. The first crops varied greatly in quality. Those of us who had been in [agricultural] school knew why, and we asked ourselves, 'What can we do to straighten this out?' And we took action. Of course, the boundary stones all had to go, and that is the difficult part now. Everyone's own land is still out there, but the markers are all gone. Not only the stones were taken out but also the paths between fields, because they weren't needed anymore, but you could tell for years after planting that it had been used as such before." According to Edgar, it took about five years for the merged fields to measure between 5 and 10 hectares and another five years for them to reach their present size of 20 to 30 hectares.

The documents of incorporation for LPG I Zur Linde were not drawn up until the harvest of the fall of 1960 had been completed. With their acceptance, the Schorchts had taken the first step into the world of socialist production. "We entered LPG I under *Freiwiliger Zwang* [voluntary coercion]," says Edgar with a smile. The newly founded collective consisted of forty members, a chairman, a five-person board of directors, and a general assembly in which everybody had a right to speak up. Wally says, "There was a lot of partying in the LPG! A calf or a sheep would be slaughtered, and we would all get together in the *Gaststätte* to eat, and afterwards there was dancing."

Under an LPG I, land alone was collectivized. Even though the freehold status of the property was not lost to the original owner, control was. The Schorcht fields, some of them constituting the dowry that Anna Magdalena Stöckel had brought to her marriage to Johann Heinrich Schorcht in 1786, were to be amalgamated with those of thirty-nine other Göttern households and worked communally by squads. Amalgamation, however, required time. Property boundaries, carefully maintained through generations of ownership, could not be readily eliminated overnight. Painful as it was for Edwin to watch the traceries of the Schorcht holdings disappear one by one, he obtained some measure of satisfaction by observing the immensity of the communal fields coming into being. Their very size promised the advent of a new collective power.

Observing the course that collectivization was taking, Edwin might well have thought from time to time of his paternal grandfather, Wilhelm, who a little fewer than a hundred years before had helped liberate the Schorcht holdings from the communal constraints of the three-field system. What would Wilhelm have thought of his great-grandson Edgar toiling once more in concert with other members of the collective, reminiscent of the way their peasant forebears had coordinated their labor in the open fields from time immemorial?

Having reached age seventy-one in 1961, Edwin kept his distance from the communal activities. Working with his horse Felix, he pulled the grain wagon and other loads and helped out with the haying. He was still in fine physical condition, carrying himself tall and straight the way he always had. Edgar says of his father, "We were surprised that he put up with so much; it was really hard for him. He used to sit here at table with us. He was still very sprightly. He was skeptical [about what was going on], and when something didn't work, he was right there to point it out. He would tell us, 'Look, you people have to pay more attention to what you are doing!' I used to tell him. 'Opa, maybe you should be our brigadier! Then everything would work out right.' He was used to doing everything in an orderly, clean, and precise manner, and when everything was done communally, some people didn't pull their weight and did things superficially, and he couldn't stand that. He got mad!"

Working Communally

It seemed strange at first to Edgar and Wally to head off in the early morning to work on what had become a combination of fields rather than one of their own, even though they knew that it did not make any difference where the work started. What mattered now was that they were all eating from the same trough as far as the fields were concerned. It was not easy, though, to substitute "we" for "I," as they had so often heard they should, or to stop worrying as to whether their own interests were being safeguarded. Yet when it came to working, they found out that they could adapt to a radically different set of motivations. Edgar says, "We were forced to join, but once we were in the system, we took it seriously. We worked just as hard as when everything was still private."

However, "just as hard" did not necessarily mean with the same sense of gratification. As Wally explains, "Before it used to be a little bit of everything. We didn't have to get so many potatoes in or weed the fields, but in the collective, those large fields, well, that killed you."

Edgar adds, "It was more interesting privately, because the whole day consisted of several kinds of work; in the cooperative, you start early in the morning and do the same kind of work all day, like a factory—like on a machine! It's so monotonous. Before, you first fed the animals, then got the feed, then the work started, then there was lunch, and then in the afternoon, depending on the weather, we did something else; in the cooperative, you are always tied to one place—you had to plow the whole day or plant or whatever."

On weekends, LPG Zur Linde was often helped by workers from its sister company, the Oberweimar Limestone Company. The workers came out from Oberweimar with a truck that proved useful during harvest or when hauling potatoes to Jena, thus freeing up a tractor for other work. Edgar says, "Without them, we would never have made it." The workers were always invited to the LPG Zur Linde end-of-year party as well as to other occasions when there was a chance for farmers and workers to socialize.[13]

Decisions as to who worked where and with whom were largely in the hands of the brigadier, albeit made with a certain amount of give-and-take. In general, the principle of "voluntary coercion" prevailed in respect to task assignments. Usually women worked together, often weeding, one next to the other down the length of the field. If all were equally adept, the results led to good crops. If not, additional work was required, with some attendant stunting of the plants.

Early on, horses were used to work the new large fields, as they were not as heavy as tractors. At any one time, as many as twenty horses would be working one field. Edgar, who had good horses, if not the best, was assigned to work in the big fields. Mowing with a grass mower usually required a team of five men. When it came to spreading manure on the fields, men and women worked together, the women loading and spreading by hand while the men drove. This was hard on the women, who had been used to working in the small private fields; now, they would work for days at a time in particular sections. Inevitably, results were uneven. One woman would spread the manure as she had properly learned to do it, while another would just throw it around at random to be done earlier. Edgar says, "There was some anger and debate at times." Wally recalls with annoyance one of the most disliked tasks assigned to the women's squads: "We had to pick stones off the fields—pick stones! The tractor drivers drove around, and we had to run after them and throw the stones up on the wagon. There were always some who tried to get around it, who didn't bend down each time!"

Used to working with only other members of the Schorcht household, if not alone, Edgar and Wally discovered that working in a team brought some advantages. As Wally says, "We all worked together and also had fun together." The coffee breaks were looked forward to, not just as a way of catching their breath but as an opportunity to laugh at one another's jokes. Wally says, "We held together a lot more then. We were forced to stick together, and although we had difficult times, we had fun for ourselves at cooperative organized social gatherings. We went on outings together in a bus or to dances. That is the way we made it through those years."

Women's Day was an occasion, as were others, for supper and a dance afterward. Edgar says, "The things we didn't have were forgotten on those days." Remembering one particular pleasurable event, Wally laughs and says, "It was my birthday. We were in the same LPG. We had just come from the fields, and we drank wine and ate pastry, and even before that we had been at a neighbor's birthday party, so that everyone was in good spirits when we got here. Then, as a gag, someone decided we all had to do somersaults here in the kitchen. It was a tough time in the cooperative, but we made the best of it."

Dancing, which both Edgar and Wally enjoyed, now became a daily part of their lives. Edgar says, "Every morning [at 5:30] we would come down here, switch on the radio, and dance around the kitchen." Adds Wally, "We had to feed the animals, but we always danced first."

"Early on," says Edgar, "there was good music." Edgar bursts into song: "*Kannst du tanzen, Johanna? Tanzen Johanna, Tanzen macht Spass*" ("Can you dance, Johanna? Dance Johanna, dancing is fun").

Having thrown in their lot with the others, Edgar and Wally now felt freer to do what pleased them. Yet their occasional lightheartedness could also be seen as compensation for the greater hardship of their daily lives. As Wally says, "We made the best of hard times." Whatever the reason, they continue to look back at those moments with pleasure.

Remuneration

Since the LPG stalls had not yet been completed, the Schorchts retained their cows in their stalls along with some that belonged to the cooperative. There were nine cows altogether: six belonging to the Schorchts, and three to the LPG. Once a day, a calculation was made as to how much milk belonged to the Schorchts and how much to the LPG, and then it was divided up. Wally did the milking by hand, with other women helping with the LPG cows on a monthly rotating basis. Milking at the end of the day was sheer drudgery. Edgar says, "Milking by hand was the hardest work we had. It does a job on your hands and your nerves. And they [the women] worked in the fields all day and then they had to come home at 11:30 and prepare a meal for everyone and then go back out again and then return, feed the cows, and still have enough strength in their hands to get the milk out." The assignment of tasks according to gender was accepted practice: women milked, men drove tractors. According to Edgar, "They gave the hardest work to the women. There were men who did nothing but sit on tractors all day—not all the men but half of them at least."

Hay was also allocated. Each farmer received his allotment after it was weighed. The Schorchts—Edwin, Edgar, and the young Erhard—picked up their consignment at the end of each day, pitchforking the hay from the wagon into the barn. When they were still private, moving hay from wagon to loft was done once, after it had been cut and gathered; now it was a daily operation and was resented as such.

At first, in their LPG I, the Schorchts were paid in animal feed. Unable to produce feed for their own stock, they depended on the LPG for their supply, which they received as compensation for their communal labor. Edgar says, "Each cow was allotted an amount of basic feed [beets, greens, hay, and straw]. The power feed, the feed needed to produce milk, was distributed according to how much milk each cow produced. That was a hard battle to get through. All of us weren't producing the same amount of milk, although we were the ones who could produce the most. That's why we did OK in the first two years. But it was a lot of work. In the LPG I, we had to hit the field at 6:30 A.M., when our feed was delivered. There was a big feed wagon that was driven around to the various farms."

As for money to live on, that came from the meat the Schorchts slaughtered and the surplus milk they produced, which they were able to sell on the open market. Their skill in breeding stood them in good stead and allowed them to make profitable use of the right, permitted to them in the

LPG I, to retain ownership and disposition of their livestock. All that was to change, though, as the time was coming for them to take the next step toward complete collectivization.

Elly's Death

A heavy blow befell the Schorcht family on February 23, 1963, with the death of Elly Schorcht. Elly had not been ill or failing; her death was the result of a simple, avoidable accident that proved to be fatal. She came down the stairs to ask Edgar and Wally to look in on Roswitha again before they went to bed. She failed to turn on the light and fell on the last step, breaking her hip. She was taken to Sophienhaus Hospital in Weimar. Complications set in—a thrombosis—and she died. She was in her seventieth year. It seemed painfully appropriate to everyone that Elly, who had given so much of herself, should die in the service of a final caring act.

The grief in the Schorcht household was numbing. Edwin, who had expected to face the uncertainties of the future with Elly at his side, summoned all the powers at his command to maintain his composure as head of the family. Edgar could not believe that his mother, whom he knew to be "the personification of love," would no longer be there to smile at him. Erhard was morose. His grandmother was, he believed, one of the few people who understood him. Norbert grieved for the Oma who cared for him while his mother and father were busy with farm work. Roswitha cried.

Edgar sent off a telegram to Irmgard: "OUR LOVING MOTHER DIED MONDAY. MORE ABOUT VISA LATER." Irmgard was distraught. She wondered whether she would ever stop crying from the sadness she felt. Even with Karl at her side, she experienced a profound loneliness. The only thing that helped was knowing she had to gather herself together to make the trip to Göttern as soon as possible.

On the following day, a registered letter arrived for her from Edgar. He wrote:

> We were just about to go into Weimar to visit our loving mother when we received the news that she had passed into eternal sleep suddenly this afternoon. She unexpectedly took a turn for the worse. Nurse Marta said that it was a blood clot. We are very sad. Father was there on Wednesday and she was still alert. I am going into Weimar now. Opa can't come with me—he is too shaken up. We want to get your visas. The mayor of Göttern has already put a call through to Weimar. I am supposed to go in on Wednesday. The memorial service will take place on Thursday.
>
> Your loving brother,
> Edgar

Fortunately, they did not send the visas for Irmgard and Karl to Pirmasens—they would never have made it in time. Instead, they cabled them to the border crossing they were going to use. The wall had already been up for two years.

Commentary

The onset of the collectivization of agricultural production in East Germany in the 1950s and 1960s was an undertaking of immense scope. When I ventured out into the countryside around Weimar and Göttern, I found myself thinking about this vast social experiment carried out at the back door of the West. Wherever one drives in the countryside in the former German Democratic Republic (GDR), two telltale reminders of that past are evident: large fields and clusters of buildings, stables, milk sheds, fodder storage barns, repair shops for farm machinery, and administrative offices—semblances of rural "factories." A melancholy aura of abandonment pervades: fading whitewash, peeling plaster, rusting equipment. Once at the core of socialist East German interest and rural economic investment, the former collective's peripherality reflects the changed agenda of a capitalist economy.

A closer examination often reveals, however, animals in the stables and equipment in use, owned by privately organized agricultural cooperatives and operated by markedly reduced staffs and overhead costs. When these conglomerations were built in the fifties, sixties, and seventies, they were put up with the earnest conviction that they were to permanently transform the East German countryside in the direction of narrowing the social and psychological distance between rural inhabitants and those of the city, thereby creating the society of the future.

Like so many of their age cohorts, Edgar and Wally Schorcht were participants in that undertaking. Initially, when LPGs began to appear here and there in the county of Weimar in the mid-1950s, and even in Göttern in the form of LPG Gemeinschaft in 1958, the attitude of peasants with middle-size farms was that the pooling of resources was an excellent solution for others—small farmers, chronically bereft of machinery and other agricultural necessities—but not appropriate for them. This was decidedly the case with the Schorchts, who as independent farmers were enjoying greater success in the late 1950s than they ever had before. Yet the political pressure exerted by the SED in the "Socialist Spring" of 1960 raised the economic penalties so high that capitulation appeared inevitable. One might think that joining under such circumstances would engender strong resentment on their part, so that they would start collective farming armed with discontent. That this was not the case is a matter of some importance.

The narrative of this chapter suggests that their initial attitude toward communalism was positive and hopeful, not resentful. When Edgar says, "We worked just as hard as when everything was still private," he implies that they were as determined to make a success out of "their" LPG as they had with their farm. Demonstrating affirmation rather than opposition may have been a hard-won psychological battle, but it was won with the understanding that pervasive negativism could only be self-defeating. Aiding them in this resolve was the knowledge that LPG Zur Linde was not so much imposed on them as largely created by them and the other *Grossbauern* of Göttern. As Edgar says, "Type I, that was the nicest time, and afterward it was never quite as good. We really came together and made an effort"; and "In the beginning it was hard, but later we got used to it, because it was necessary to work in better ways." "Necessary" here refers to the economy of size that the larger fields were perceived as providing.

It was not that Edgar and Wally preferred working communally; in fact, they found it more monotonous. Yet laboring in a brigade brought certain advantages, more psychological than instrumental. The brigade was conducive to having fun, "fooling around," telling stories, cracking jokes, and letting off steam. Edgar and Wally remark often about the good times they had in LPG Zur Linde at work and at play. But this pleasure was not had at the expense of the task at hand.

What I have written here in August 1995 finds parallels in a *New York Times* article appearing on October 31, 1994, in the Business Day section. Under the headline "G.M. Success in an Unlikely Place," James Bennet states that the Opel automobile plant in Eisenach, East Germany, has found that the success of the plant has been achieved "not in spite of, but because of, [the workers'] experience and lingering ideals." He writes, "The East Germans, G.M. discovered, are delighted to work in small teams, like the 'brigades' they knew in the Communist days. And they are happy to strengthen their bonds by socializing after hours, at the company's urging, with fellow team members."[14]

The pride with which Edgar refers to LPG Zur Linde carries the same ring of conviction that characterizes his praise of the Schorcht farm at its peak before collectivitization. Although Edgar and Wally never succeeded in substituting the collective "we" for the individual "I" in their thinking, they were able to find in the realities of collectivism satisfactions to which they still give credence in their reflections on that time.

Edgar Becomes Mayor

Collectivization, LPG III

Edgar Returns to School

IN 1964, A YEAR AFTER his mother's death, Edgar, wishing to become the brigadier in charge of animal breeding in the agricultural production cooperative (LPG), applied to enroll in a master class in animal husbandry. Initially, it would be the responsibility of the brigadier to go from farm to farm to coordinate the breeding; later, he would direct the communal stable to be built in Magdala. The decision to return to school with pay was not, however, Edgar's to make; it depended on a vote of the LPG membership. The balloting was seventeen against and sixteen in favor of his request. Edgar remains convinced that the vote reflected the chairman's antipathy for him. Edgar says, "Our leader here didn't want me to have this training, because he thought I would try to take away his job. . . . But I attended the classes anyway and just wasn't paid for the days I was in class."

Despite the loss in credit for animal feed that Edgar's schooling cost the farm, amounting to between 3,000 and 4,000 deutsche marks (DM), the Schorchts were still able to make ends meet due to the milk-production money. What galled Edgar was that the chairman attended the classes, too, and was paid for his time in the classroom.

The program took place in Magdala, with classes meeting two to three times a week from fall to spring. Edgar attended for four winter terms. Because he had frequented school only until the eighth grade, he had to make up the ninth and tenth with a full load in German, math, and physics. At age forty-five, he graduated with distinction.

Edgar Becomes Mayor

In 1951, Edgar became a member of the Liberal Democratic Party of Germany (LDPD). The LDPD was founded on July 5, 1945, roughly one month after the Soviet authorities permitted the formation of political parties and trade unions in their zone of occupation. It was undoubtedly the intent of Article 8 of the party's program that held the most interest for the Schorchts, especially for Edwin: "the retention of a unified German national economy, of private property, and of the free market is a prerequisite for initiative and for successful economic activity."[1]

Edwin had become an LDPD member in 1949, as had other farmers like him, convinced that with the LDPD's commitment to a market economy, the party stood as a bulwark against the collectivist goals of the Communists. Edgar reports that his father often said to him, "You must join, too, so that we can stay strong against the Communists." Edgar says, "It was the party of teachers, artisans, and farmers. It was the party that thought something could be saved after the war.

"They were looking for members and they came after me, but I never went to their meetings. The mayor for our town always came from the LDPD. That is the way it was: one town had a mayor from the CDU [Christian Democratic Union], another from the LDPD, another from the Farmer's Party [*Demokratische Bauernpartei Deutschlands*, or DBD]. When it came time to change mayors, the party itself had to put up another candidate for the spot. If it couldn't, the 'Big Brother' party [the Socialist Unity Party of Germany, or SED] offered someone for the position, and they always had someone ready. We had a mayor, Erich Bertisch from Weimar. He was here only three to four years and just came for 'visits,' you could say. He was relieved of his position, and they asked me if I could do the job. I said I could, but I didn't want to.

"[At the same time,] one of my teachers from Weimar, Erich Pfeiffer [from agricultural school], came and told me to take my state exam in agriculture in what was later called agricultural engineering. I told him I was interested but that I had to put in my hours at the LPG to get feed for my animals. If I wasn't there, then we had to get the feed some other way. The people from the [LDPD] council kept coming almost every day to ask me to be *Bürgermeister*. I sent them all away. They came back the next day. I told them I was going to take my state exam. They said, 'No problem—we'll delegate you to do your state exam as mayor!' The next time they came I thought, maybe I'm doing the wrong thing if I don't, so I gave in and conceded—that was in October 1965. I became mayor." Edgar remained *Bürgermeister* for twenty-two years.

Edgar as Bricoleur

In the beginning, Edgar worked in the LPG in the morning, served as mayor in the afternoon, and continued his training as well. He finished his course work for the state exam in 1967, which involved, among other things, attending milking school in the village of Gutmannshausen, north of Weimar. Edgar says, "That's where we really learned how to milk, by hand and mechanically. It was theory and practical experience as well. We had to go out and milk early and then again in the afternoon. It was nice out there, but my wife was all alone, stuck with all the work on the farm."

Special attention was given in Gutmannshausen to mastering the operation and maintenance of the large mechanical milking systems that were beginning to be installed in LPG cow stables. These systems involved the extensive use of glass piping, and everyone who milked was supposed to know how to repair breaks in the glass lines.

Edgar says, "At Gutmannshausen, we learned how to repair everything, and that's the way it was in the GDR [German Democratic Republic]. We became masters of all trades. When my brother-in-law came from the West in his Mercedes, he would scratch his head and cuss, but he never bothered to look inside. The main thing was that it ran and he knew how to drive it. But me, if I drive it, I have to know how to take it apart and put it back together again. That's the way it was with the Bulldog [tractor]. If something was broken, we would go to Grossschwabhausen, where there was a garage, and we would watch them take everything apart, clean everything out, and put it back together again. After that, we knew how to do that ourselves [*laughs*]. But no one is learning how to 'do it yourself' anymore. That way of doing things is all going by the boards."

Transfer of the Farm from Edwin to Edgar

Although Edgar had been operating the farm since 1957, it was agreed between father and son that with the completion of Edgar's training in 1967, the time had come to transfer the property formally from Edwin to Edgar. Edwin's stewardship had lasted forty-five years. Though seventy-eight years old, he was still hale and hearty, and busy—grafting fruit trees, clearing a woodlot, harnessing Felix for a day's work. It was circumstances that suggested the time was right.

Ten days before Christmas, Edwin and Edgar presented themselves before a notary in Weimar, Hans-Georg Schmidt, who had already drawn up the document on the basis of prior consultation. On their way to Weimar, Edwin reminded Edgar of how he had traveled in company with Edgar's grandfather, his three aunts, and their husbands in the buggy to

Blankenhain forty-five years before—the year Edgar was born. They had, in a sense, come full circle. That they were going to Weimar instead, the new seat of jurisdiction for such matters, seemed to be in keeping with all the other changes that had occurred since those earlier days.

The concise typewritten three-page document begins as does its predecessor of 1922, listing the same seventeen separate fields and woodlots that were to be transferred.[2] The difference between 1967 and 1922 was that the LPG had made the physical boundaries of the fields indistinguishable by amalgamating them with others, even though the land-registry book numbers and field names remained individuated. The tie, existing for more than a hundred years, between the Schorcht name and those lands had been broken.

The substance of the remainder of the document has to do with Edgar guaranteeing to his father lifelong lodging and boarding, on the first floor; full use of the bedroom that looked out onto the courtyard, including heat and light; and use of the kitchen, living room, work space, and house garden. Unlike Edwin had been, Edgar is not responsible for his father's clothing. Gone in the 1967 document is also any reference to the older man's medical treatment, since he was already enrolled in the GDR state-run health plan. Nor is Edgar expected to provide spending money equal to a hundredweight of rye, as Edwin had to do for his father. The difference may have been as much a question of Edwin's prideful insistence on independence as a formal change in expectations in regard to filial responsibilities.

One reason why Edgar had postponed the transfer as long as he had was that although he had two children, only one could inherit from him. He dreaded having to come to grips with that inequity. Waiting had assured him that Irmgard was not a single woman struggling to make a living but "well married." Karl Stüber was a man of some substance; it was clear that Irmgard would be well provided for by her husband.

Edwin had been advised by the notary to come to an agreement with his daughter, lest the ownership agreement become unduly encumbered. Irmgard was convinced in her own mind that the document should bear only one name as the owner, that of her brother, even though she expressed some feelings of hurt at not receiving anything from the property. Edwin tried to provide his daughter's "share" by other means. Free to travel to the West as a retiree, he went every year to Pirmasens, smuggling cash across the border, sewn into his tie or underwear. For a man who held the law in high respect, even that made by the Communists, breaking it would have been intolerable if it had been for any other reason. It was always disconcerting, though, to find how 1,000 hard-earned ostmarks melted away in the exchange to a mere 200 DM. Irmgard could only find it a little ludicrous that her father once arrived with valuable goose down sewn inside his winter coat.

LPG III and How It Worked

In the late 1960s, the SED mounted a major campaign to induce farmers to join type III LPGs in keeping with the long-term goal of moving toward the collectivization of all agricultural resources—in effect, cutting all remaining ties with independent farming. It required the farmer to bring into the pool not only his land, valued at 500 DM per hectare, but also his livestock and equipment. Contributions in excess of the required entry level were credited to each farmer, with the balance to be refunded later.

All the earnings made by the LPG unit in the course of the year belonged to the members. A good portion, either one-third or at least one-fourth, was set aside in a general fund for acquisitions deemed necessary. The remainder of the earnings was distributed to the members, each being paid according not to his original contribution but to the perceived value of his or her work.

The productivity of individual members was calculated in terms of work units (*Arbeitseinheiten*, or AE), so that one eight-hour day of light work in the field was worth 1 AE, with heavy work earning 1.5 or 2 AE. Different ratios were established for tractor drivers, technicians, and persons with particular skills or responsibilities.

The collective made use of the general funds for acquiring new machinery and breeding stock; erecting farm buildings; and building community centers, kindergartens, and sports facilities. Contributions were also made toward sick pay, paid holidays, and pensions. It was expected that LPG members would work an eight-hour day for five days of the week.

Precise guidelines for the operation of the LPG came from the state, but decisions as to the actual conduct of the collective were made by the full assembly of members who elected its executive committee, chairman, and other responsible officials, including the brigadier. The brigadier oversaw much of the day-to-day administration of the collective, including making decisions assessing the productivity of each member.[3]

By 1975, the number of type III collectives had increased dramatically, numbering 93 percent of the GDR's collective farms, while the number of type I and II collectives decreased between 1964 and 1975, from 9,566 to a mere 306.[4]

The Schorchts Enter LPG III

Edgar and Wally would have preferred to continue in the LPG I, retaining control of their livestock and equipment, but the cards were stacked against them. They understood that for large farmers like them, the economic penalties for staying out would be heavier than they could bear. They were liable to threats as well, such as having no fertilizer or diesel fuel.

When the time came, it was hard to see the equipment being rolled away. The cutter, the plows, the wagons, and the Bulldog were taken, making for large empty spaces in the barn, now at the disposal of the LPG. The cows remained, even though they belonged to the collective. Indeed, more cows were added to those the Schorchts were already stabling. They had to accept an additional nine, crowded into what had been the young-stock stall. Never had the courtyard seemed so narrow. Wally had more work to do than ever, caring for the additional animals, working in the fields, and returning to the house to prepare the noonday meal.

Joining an LPG III superseded the prior contractual arrangements the Schorchts had made with the LPG I, meaning that calculations of starting costs were undertaken with a clean slate. Each farmer was obligated to "bring in" land as well as animals and machinery. The average farm size in Göttern was figured to be 14 hectares; thus, for any amount of land up to that amount, a member had to bring in 500 DM per hectare in goods—that is, livestock or equipment—land being regarded as a debit, since it had to be worked. For hectarage above that, the farmer was required to bring in 1,000 DM in goods. Accordingly, the "start-up" cost for the Schorchts was 21,000 DM.

The cattle were evaluated by the Breeders Association for premium amounts due to their high quality. Their best cow was appraised at 2,200 DM, twice the average amount, and the remainder between 1,600 and 1,800 DM per cow. The tractor was estimated at 2,900 DM, while the field wagons and tools were assessed minimally. The value of their inventory exceeded the amount they were required to bring by 18,000 DM, with 25 percent of this "inventory surplus contribution" to be paid back without interest over a thirty-year period.

The strength of the Schorcht farm made these "entrance costs" onerous but affordable, allowing them to join LPG III in good standing. Not so fortunate were others who could not come up with the start-up costs and so entered with debt to be deducted from earnings rather than credit to be repaid. Unfortunate were those who came in late, as was the case with the family of Edgar and Wally's daughter-in-law in Milda. When they applied to join after everyone else had, the LPG told them, "We're doing just fine without you, but we will take you if you pay 3,000 DM per hectare!" They had to sell their barn to join. Edgar says, "When we turned over all our assets back then, whoever thought *der Wende* [the change] would ever come!"

Now, for the first time in their lives, Edgar and Wally were paid employees. In the beginning, they received the equivalent of 1.80 DM per hour. Later, it went up to 2.80 DM, although some categories of workers received as much as 4 DM. The average worker's pay was 800 DM per month, but others— such as combine drivers, who often worked late into the night—earned as much as 1,200 to 1,500 DM.[5] A family's well-being no longer depended on the productivity of the farm but on the success of the collective. Edgar took pride in their LPG. It might not have been the best; as Edgar admits, "The best LPG was Ossmannstedt, but they also had the best soil. We were pretty good, though." The substitution of "we" for "I" began to assume a new reality. Edgar says, "We thought that it would just go on like that forever."

They were still able to slaughter, allowing them to make some extra money by selling on the open market in Jena (a 13-kilometer trip) or elsewhere. It was a risk, however, for their ability to sell depended on whether the people they were selling to were paid on time. The day Edgar and Wally went as far as Kahla (25 kilometers away) on a Monday—payday at the local porcelain factory—the workers had not been paid. Unable to sell, the Schorchts left the meat from two of their pigs with the butcher.

Edgar and Wally's First Vacation

From 1965 to 1971, while cattle were still stabled in their own stalls on the farm, Edgar scheduled his work in the mayor's office so that he could help with the feed in the morning and then with cleanup in the evening.

One of the first things Edgar did when he became mayor was to join the Mayor's Association. Although it was expected that he should join, an additional incentive was two-week vacation trips offered to members. One could choose to go to the Thuringian Forest, Berlin, Vogtland, or the Harz Mountains for only 85 DM for the member and 125 DM for the spouse, everything included. It was also possible to travel to the Soviet Union or other Warsaw Pact countries.

In 1970, Edgar and Wally took their first vacation. The fact that Edwin, a disciple of duty and hard work, had allowed himself and Elly to go on vacations before the war made it easier for his son to follow suit. Edgar says, "Lots of people in the LPG never took vacations. They [felt they] didn't have time to rest. But we said, 'Why shouldn't we go?' So we went to Russia in 1970. We paid for that ourselves, 520 DM, all meals and pocket money included. We went by airplane, eight days long."

Wally had read an ad in the paper advertising a one-week trip to Kiev and Leningrad, saying to Edgar, "Didn't you always say that you wanted to go back to Kiev?" So they signed up for the trip, obtaining two of the last spots.

It was Wally's first major trip. In preparation, they took an hour-long test flight, leaving from Erfurt, flying over Göttern, picking up Norbert and Roswitha in Jena, and then returning to Erfurt. Neither the trial flight nor the one to Kiev prepared Wally for the bus ride from the Kiev airport into the city. It left her shaken. On the way in, Edgar noted the same wooden houses with the straw roofs that he remembered from the war. What was startlingly different were the blocks and blocks of new apartment houses, made of prefabricated box construction, similar to the ones they had seen going up in Jena and Weimar.

In Kiev, they stayed at the Opera Hotel, where they fell into conversation with two Russian officers who had been stationed in Potsdam, Edgar having remembered a few words of Russian from the war. One of the officers told Edgar and Wally how much they were enjoying the conversation and that when they were stationed in Potsdam, they had not been allowed to speak or fraternize with Germans.

Before long, Edgar recalls, a bottle of vodka, four glasses, and a plate of meatballs appeared. It being noon and Edgar unaccustomed to hard liquor at any time, Edgar was thankful for the meatballs, which allowed him to stay glass-to-glass with his hosts until the bottle was emptied.

That evening, there was a wedding reception at the hotel, to which the Schorchts were invited. Wally recounts with a broad smile how she danced vigorously with the groom and then observed with amazement how often people hugged and kissed each other, even the men. Wally says, "In the GDR, men would not do that, except [Erich] Honecker and other big shots, copying the Russians."

Edgar was surprised at how many Russians spoke German, but he also noticed a hesitancy, as if they were still afraid of them—an impression, he says, that was confirmed when they visited the Piskaryovskoye Memorial Cemetery in St. Petersburg, where the guide never mentioned the Germans by name, speaking only of "the Fascists." Edgar said that he was not prepared for what he saw at Piskaryovskoye: block upon block of mass graves of the people of Leningrad, "after we Germans had besieged the city for nine hundred days and driven the people into starvation. I felt ashamed as a German"—Edgar stopped speaking, looked away, his eyes filling with tears.

The Magdala Stables

By 1971, the cows had been moved out of the Schorcht stalls to the new ones in Magdala. Edgar says, "It was hard to adapt. We would have to get milk from Magdala, and then later we had to buy it at the *Konsum* [cooperative]!" With the cows gone, Wally stopped working with the field squads and was assigned exclusively to the new collective cow stalls in Magdala. All of Wally's co-workers were women, a fact that did not surprise her. It had become the norm in the LPG that handling cattle was women's work; it was assumed that they were "by nature" best suited to such responsibility.

The day the cows left their barn in Göttern was a sad one for the Schorchts. Edgar says, "There were more than just a few tears that were shed—not just us, the animals, too. It took a long time until they got used to being somewhere else," Edgar continues with a grin. "They had to become socialist cows. The milk production stayed the same for a while, then slowly went down."

The change took its toll on the women as well. On the one hand, the milking machines made for less moving about, but on the other hand, one still had to go home to prepare the noonday meals, as well as get up at four in the morning to arrive at the stables on time. Wally remembers there being women in Bucha who worked two shifts a day: "They were always so totally exhausted, those women, always so pale; they never got out in the sun."

Each stable in Magdala held 216 cows. Milking was done by machine, with the milk running through glass tubes to a 5,000-liter storage tank. Wally considers milking the hardest work she had to do at Magdala, despite the machines they used. Wally serviced thirty to forty cows, and each udder had to be washed before milking. She had to bend down three times to service each cow: "It was very bad for the back; that first year, my knees hurt all the time."

After washing the udder, a first sample was taken by hand to check whether the milk was "clean." The milking device was hooked up, and Wally would check to see whether the milk was running properly through the tubes. By the time she had prepared the second cow, the first was almost finished.

Each milking took between six and eight minutes. Wally then had to "post-massage" the udder to make sure that all the milk had come out. The last milk to leave the udder contained the most fat, which was important, as the women's ratings were based on the milk's fat content—they aimed at 5 percent rather than 3 percent. When no more milk would come out, Wally detached the device from the cow's udder and did the last milking by hand to get the final drop.

At the end of the milking, Wally would take a pitchfork and separate the clean straw from the manure. This had to be done before the tractor came through to push the manure out of the stable.

It was a relief to be assigned, from time to time, to the birthing stall. The cows to be delivered were moved in a week before calving. Eight weeks prior to delivery, they were started on a better feed of wheat mixed with soybean, sorghum, barley, or oats, so that the calves would get off to a good start. Two women were in attendance at delivery.

As time went on, there was always less need for Edgar in the LPG. After ten years, he stopped participating entirely, giving all his energy to his duties as mayor. In the last few years, he helped with the young stock or worked in the fields getting the potatoes in. When the task forces of factory workers came to help their fellow laborers in the countryside at harvest time, Edgar would show them where and how to work in the fields. According to Edgar, more than a few of these workers did not like to give their all to what they were doing, kneeling and bending as if they had a disability.

From LPG Zur Linde to
LPG Hermann Matern

With the building of the stables, milk rooms, feed sheds, and personnel facilities in Magdala in 1971, LPG type III Zur Linde joined with the collectives of Magdala and neighboring Ottstedt to form a considerably larger unit, LPG type III Hermann Matern, named after one of the founding figures of the land-reform program.

The basic organizational structure of the Hermann Matern collective was essentially the same as that of the smaller one: theoretically, all decisions were made by the assembly, then implemented by the chairman and the brigadier, and each member had a vote. The democratic promise underlying the organization was not, however, always fulfilled, and even less so as time went on. At first the meetings were held every month, eventually tapering off to just twice per year, with fewer votes taken by the assembly. Finally, a meeting was held once, at the end of every year, to plan for the coming year.

Management, which was supposed to be compensated according to published guidelines, set its own wage scale. Gains that members believed they had made by improving performance were lost because of the poor performance of the leadership. Edgar says, "If they didn't meet their quotas, then it should have come about that their pay was docked. But that rarely happened, and when it did, it hardly ever hurt because they earned so much."

The social occasions, if anything, increased in frequency and elaborateness. Wine and dinner evenings were sponsored and paid for by the collective, although contributions were exacted from everyone. The spirit of commitment that had prevailed among members in the sixties and seventies began to give way to indifference in the eighties. Edgar says, "People had lost the feeling that what they were doing was for their own benefit. They began to say things like, 'Brigadier, if we go out at breakfast time, that will certainly be early enough!' In the seventies people still worked hard and well, but in the eighties there were always some who wouldn't work at all."

In Edgar's opinion, if collectivization had remained as it had been with the Göttern LPG, it would have succeeded in the long run. He says, "With Göttern and with Magdala it still worked, and it could have stayed that way for the new and larger LPG if we would have been even more productive. But 'they' said no; we have to make better use of the machinery."

The Metamorphosis from LPG to KAP

By the early 1970s, it was clear that comprehensive collectivization was not to be the final phase of the reorganization of agriculture. The GDR leadership decided to move beyond collectivization toward the industrialization of the land. It was held that an industrialized agricultural system would improve productivity through the creation of larger and more specialized units, further overcoming the disparity between field and factory. In the early 1970s, the *Kooperative Abteilung Pflanzenproduktion* (Cooperative Department of Crop Production, or KAP), was formed by merging existing LPGs. By 1975, approximately 1,200 KAPs had been established in the GDR, accounting for 85 percent of agricultural land.[6]

In the Weimar district, the number of LPGs dropped from fifty-nine to thirty-three in 1975, as KAPs were formed in the areas of Blankenhain, Hochdorf, Bad Berka, and Tonndorf.[7] Opting for an economy of size, each KAP covered several thousand hectares, including all the ancillary facilities involved in plant production. In 1978, the Rote Banner (Red Banner) LPG of Niedersynderstedt joined the LPGs from Kesslar, Magdala, Göttern, and Ottstedt to form a KAP that encompassed 3,400 hectares and stretched over an area of 10 square kilometers. "But by then," Edgar says, "people weren't interested anymore. People started saying, as they hadn't before, 'This is what used to be ours . . . ,' because they kept adding and changing things all the time." The substitution of "we" for "I" began to lose conviction.

Edgar and Wally's Silver Wedding Anniversary

In 1973, Edgar and Wally celebrated their silver wedding anniversary (*Silberhochzeit*). There had not been a celebration like this in the family for some time. Edwin and Elly's silver anniversary had arrived on June 28, 1945, just a short while after Edgar had returned from his internment. It did not seem fitting to celebrate then after the war and so much sadness. Wally's parents had not observed their silver anniversary, either, out of respect to Reinhardt, still missing in action.

For a week beforehand, Wally and her mother baked while Wally kept up with her responsibilities in the stables. She knew she would be exhausted when the guests arrived. With Edgar as mayor, they had many obligations. Edgar says, "So many guests! First of all, thirteen mayors from the Community Association right after breakfast, then the village council from Göttern and the fire department, and members of the Weimar District Council and the Party Council. There were eighty people or more, and some of them stayed until 6:00 the next morning! There were always some more who wanted to come and congratulate us, and you have to be prepared for those people, too." For the occasion, Wally bought a new dress and a tiara to adorn her hair. She was complimented on how smart she looked.

Edwin's Death

Edgar remembers it was the day he attended a Community Association meeting in Kranichfeld when Wally called to say that Edwin had died in his sleep. It was September 23, 1976. Edwin was eighty-seven years old.

Until he turned eighty-five, Edwin was still keeping busy with a few select chores. He did some pruning of the fruit trees and picked apples in the fall. That winter, he had begun to weaken noticeably, finding it increasingly difficult to walk any significant distance. But when summer came and he turned eighty-six, he set his heart on making his annual trip to Pirmasens to visit Irmgard. He especially wanted to be there for her birthday on May 22.

Since he had become a pensioner in 1954 at age sixty-five, he had been granted the right to visit a family member in the West for as long as twenty-eight days a year.[8] Edwin, together with Elly when she was still alive, had always taken advantage of this privilege, thoroughly enjoying their stay in the West, where everything seemed so different and up-to-date.

When he returned from Pirmasens, he seemed rejuvenated. The summer passed, and the fall approached. To everyone's amazement, Edwin held his ground. Shortly before he died, he confided to Wally that he hoped to be around for his birthday in October.

As soon as he returned home, Edgar called Irmgard. He then made a trip to the police station, where a telegram was readied for the post office authorizing Irmgard's visa. The body was brought to the crematorium in Weimar. Since Elly had been buried thirteen years before, expectations had begun to change in regard to how to properly dispose of the dead. Now it was customary to cremate and to avoid the extra labor necessary to dig a grave large enough for a coffin.

After the chapel service, everyone gathered in the *Gaststätte* of the *Weimarhalle* to reminisce. The feeling was widespread that with Edwin's death, an era had ended.

Irmgard Travels East

After her father's death in 1976, Irmgard made an effort to visit her family in Göttern more often. She and Karl came east for Norbert's wedding in May 1977, and then again for Roswitha's wedding in June 1978. In 1979, they returned, their car packed with 25 cubic meters of bathroom tiles for renovations that Edgar and Erhard were undertaking. In June 1980, they packed the car again with tiles, this time for Roswitha's kitchen. For the most part, they succeeded in bringing in or out what they had with them.

The border crossing was traumatic for Irmgard, particularly when she came alone by train. She says, "The fear that I experienced there . . . they always checked us! Even something small, a wurst—I wasn't allowed to bring that. Once they caught me with a goose. They passed over everyone else and checked me and found a goose! The young man wanted to throw me off the train! I said to him. 'Look here, I was there for fourteen days and helped morning, noon, and night, and my brother gave me that. When I look at you, you could be my brother! You want to detain me because of a goose? [*Sighs.*] The fear I had!"

Finally, Irmgard was not allowed to visit anymore or to exit the Autobahn at Magdala. So she and Karl would stop on the shoulder of the road and wait for the family to walk up from the village to visit. Once they were aware that pictures were being taken of them. Later Irmgard arranged to be invited to the Leipzig trade fair. Edgar and Wally would drive up in the Trabi, and they would make a day of it at the fairgrounds.

Wally Retires

In 1981, at age fifty-five, Wally retired from active service in the LPG. The years of tending cattle and working in the fields had taken their toll, and she was able to retire on a disability pension. Wally now has to use a cane some of the time. She says, "My back is gone, my discs are pushing up on my nerves, and my whole left side is numb—no feeling. It's terrible. I can't stand for long, and no doctor can help me." Even so, she admitted to feeling sad the day she left the stables in Magdala and aggrieved for the friends with whom she had worked in the fields. She says, "They went off to the fields at noon. I stood in the gate and looked at the people to see who was going out there, and I cried."

Wally's daily routine changed drastically, and it was difficult for her to make the adjustment. For a year, she worked in the kindergarten in Magdala on a temporary basis, but she was ill much of the time, and that affected her passion. Her daily trip to the kindergarten took her by the stables. She says, "It hurt just to see them. Even though it was hard work, I liked [the people] a lot."

Now that she was retired, Wally was allowed to travel to the West. She made her first trip to Pirmasens in 1983. Edgar could not go because of his duties as mayor. Wally had never traveled so far by herself. When she arrived, many things seemed different from what she remembered from her last visit in 1961. Irmgard was living in the house that she and Karl had bought on a hill in the better part of Pirmasens. From there, Wally enjoyed going downtown to look at the shops, even though she felt nervous at being alone and seeing a *Neger* (Negro) for the first time.

As for the rich West, with all its goods not available in the East, Wally says, "I am not the kind of person who needs all the things that everybody else has, so it didn't bother me too much. I only wanted to bring some things back home with me for the others. I bought four sweaters for 22 DM for the grandchildren. They were made in Apolda [near Göttern], but we were never allowed to buy them there; the firsthand quality [merchandise] would go directly over to the West."

With the animals gone except for a few pigs, rabbits, and hens, there was no more need to keep the manure pile in the courtyard within easy reach of the stalls. Edgar saw to moving it to the back of the barn. It was a blessing to have fewer flies and wasps buzzing about the courtyard on warm summer days. When the manure went behind the barn, the chickens followed, so the courtyard, free of the hens and their droppings, became much easier to keep clean. In the summer, the family began to take to sitting out in the cool of the early evening after the day's work was through. Little by little, the appurtenances of farm life were being discarded.

Commentary

With Wally's retirement in 1981, the Schorchts were effectively out of the LPG. For some ten years of brigade work in the fields, and then for an additional eleven working exclusively in the stables in Magdala, the LPG commitment had occupied the larger part of Wally's day. Her retirement left a void that was hard to fill. Even though there were endless tasks to do around the house, it was not the same as having a job. While it was gratifying to keep the house in order, it was lonely work. It did not give her the chance to let off emotional steam, as working close to co-workers in the fields or stable had. She found herself envying Edgar. His job as mayor kept him in touch with people the whole day.

It seems that the LPG had left its mark on Wally and her expectations. For the last twenty-one years of her life, she had, in effect, worked as an employee of a state-owned agricultural industry, with the responsibilities and rights pertaining thereto: regular hours of work, alternative weekends free, vacations, health care, and a pension at termination. Like Edgar, she had grown to appreciate the privileges of employment. It would have been difficult, the Schorchts admit, to return to private farming after their experiences away from it. With new expectations developing in her thinking and conditions of her employment increasingly simulating those of her urban counterparts, Wally had quite unwittingly traveled part of the journey foreseen for her by GDR ideologues.

Unlike her sister-in-law Irmgard, Wally does not wear makeup. The lines on her face are discernible. Wally wears comfortable clothes, sturdy shoes, and a scarf tied under her chin when she goes out in inclement weather. Irmgard wears city dresses, suits, and a black broad-brimmed hat, set at a jaunty angle. Alighting from her new four-door white Mercedes, Irmgard presents an elegant figure. Yet I think it is true to say that despite her infirmity, Wally would not trade places with her sister-in-law's comfortable life in the West. In the same vein, Irmgard would not choose to return to Göttern to live. I have the impression that while she has a large warm spot in her heart for the family, her annual summer visit is sufficient for her.

Irmgard and Wally appear to have made peace with their choices. Yet one must not dismiss too readily the sacrifices they made in the service of womanhood. Irmgard was asked time and again to renounce her own wishes for what she wanted to make of herself, because such desires were regarded as unworthy, impractical, or selfish by the very people she loved the most. The social history of women is replete with instances of those whose sense of self has been disconfirmed in the name of values that serve men well.[9] For persons of lesser mettle than Irmgard, such rebuttals might have cost them their mental health. Instead, she was able to make the most of what a dismembered Germany was able to offer her: another Germany, whose distance from the other allowed her to hear and respond more accurately to her inner voice.

Wally's trial was less specific to person and family and more contingent on those larger social landscapes that have always conferred upon women second place. Saddened by her brother Reinhardt's loss but securely in love with her young farmer, Edgar, Wally chose for herself a life of hard work for which, as a woman, she received the lesser recognition. That the work became more demanding as she grew older, weakening her health, is not to be wondered at. That such an outcome did not leave her self-pitying is a tribute to her hard-won sense of self-esteem.

13

The Mayoralty of Edgar Schorcht, 1965–1987

How Edgar Became Mayor without Being Elected

WHEN EDGAR SCHORCHT accepted the position as mayor of Göttern in October 1965, he replaced Erich Bertisch, a resident of Weimar and a member of the Liberal Democratic Party of Germany (LDPD). Edgar regarded Bertisch more as a visitor than as a member of the community. For his part, Bertisch believed he was doing the party a favor by accepting the mayorship to prevent it from falling into the hands of the Socialist Unity Party of Germany (SED). In the eyes of local and county party leaders, however, Bertisch was deemed incompetent. Afraid of the harm that he could render the reputation of the LDPD, the large farmers removed him from office. It was for this reason that they had importuned Edgar to become mayor. Although not viewed in LDPD circles as a strong party supporter, Edgar was held in respect by all in the community, as was his father.

Edgar's assumption of the office did not require the legitimization of an additional election. His right to it had already been won in the voting of 1952, in which the LDPD, winning in Göttern, had achieved the mandate to name eligible mayors from its party.[1] The LDPD would continue to hold this mandate as long as it was able to present acceptable party members for the position.[2] If it was unable to do so, the office of mayor would revert to the SED. In similar fashion, Weimar had a mandate for Christian Democratic Union (CDU) mayors. The great majority of German Democratic Republic (GDR) mayors were SED party members. "Deviant" (non-SED) mayors were clearly in the minority, with no more than twelve LDPD mayors in what had once been Thuringia.[3] In some circles, they were referred to as "exotics," mayors who had greater freedom to maneuver within a strictly defined political and administrative space.[4] All of these considerations were apparent to Edgar when he said yes to the loyal band of local party leaders. In addition, he was aware that he would be automatically eligible for "reelection" as long as he and the party were determined that he should serve. It would mean the end of his career as a farmer, if not that of the Schorchts as a farming family in Göttern.

"Politics Destroys It All"

In Edgar's view, one reason why the LDPD existed was to serve as an alternative to the SED. Edgar says, "The LDPD was the party of the farmers and teachers. There were probably about seven, eight, maybe ten large farmers in it. It was a small party, but it had the best farmers, while in the Communist Party, there was almost nobody, and so that is why the mayor came from the LDPD! We in the LDPD wanted to show the others that we were better, and maybe we actually were, although that is no longer the case today. Back then, in the LDPD, we had one meeting every month with the mayors and the county representatives. There, it was decided what would be done and what would be the politics of the future so that we could all know it ahead of time. The SED didn't do that, but we did it, and did it in public, and even though there were never more than fifteen to twenty people there, everybody's work was being openly assessed. That procedure was positive for us, because if something negative came up, we would put in some extra work to fix it. Of course, among ourselves, we criticized the SED, but we didn't do that in public!"

There were some advantages in being a minority-party mayor. SED members were constantly at risk of falling into doctrinal traps. Even though he was free of those constraints, Edgar had to deal with the SED on a daily basis in the person of *Genosse* Rudolf Werner.[5] Edgar says, "He was a member of the SED county leadership. Those fellows were always so tough. Werner wanted to introduce discipline here as well. As long as he attended the meetings of the assembly here, I didn't dare hold one without giving a preamble, something copied from the *Neues Deutschland* [the SED party newspaper], some nonsense I never believed in but had to say anyway. I didn't agree with all those things—I only cared for the welfare of the town. In my opinion, politics destroys it all. But Werner was chairman of the National Front, and that gave him a lot of authority.[6] The National Front gave out all kinds of things, including awards. When we handed in files to them showing what people had done here, Göttern was always among the achievers. And that was because Werner and I always worked together. He did a lot for the village, but because of the fact that he was an SED man, I could never have neutral feelings about him. I always had to be careful, even though we worked together well."

Democratic Centralism

From early on, Edgar learned to heed directives emanating from above in the form of the County Council even while he was subject to pressures from Göttern's own council and assembly. In his opinion, neither could be denied at the expense of the other; to question the authority above would jeopardize his mayoralty—with respect to that, there were no options. Even with the full support of his party, he could not prevail against the SED, yet turning a deaf ear to his own constituents would lose him their respect, his own self-esteem, and thus his effectiveness. In short, Edgar had to learn intuitively how to function as mayor within the context of the principle of *demokratischer Zentralismus* (democratic centralism), which since 1957 constituted the operative principle of the GDR governance.[7] Assuring the flow of decision making from top to bottom throughout the state apparatus, the practice of democratic centralism left little room for self-determination on the part of the community.

The Art of "Balancing"

In the beginning, Edgar attempted to continue working half days at the LPG, getting out at 4:00 in the morning and tending to the young stock, the eighty head that Erhard and Norbert had helped him control with an electric fence. The other half of his days was spent in the mayor's office. His pay as mayor, 470 deutsche marks (DM) a month, was just a little more than half of what he received in the LPG. Edgar was convinced that if Bertisch could perform the office by "paying visits" to Göttern from Weimar, he could do as well, if not better, by being on hand at least half of every day. He discovered, however, that the demands made on him increased year by year. That may have been because he was so accessible: "I always got along well with most people," he says, "becoming the person to whom the citizens of Göttern went with their problems.

"Everything was left for the mayor—not just the municipal matters but everything. I became," he says with a laugh, "the *Mädchen für alles*, the 'girl for everything.' Every week that I went to Weimar, there was always something going on, so I had to meet with the different authorities. Someone from Göttern wanted this, someone else wanted that. At first I went by bus; later on, I bought our first car, a used DKW, which made it easier."

In the course of the day, Edgar would call on one or all of the departments of the Local Supply Economy, Trade and Supply, the Finance Committee, and the Planning Commission to make requests on the behalf of individual citizens of Göttern for "balanced materials."[8] Most of these materials (wood, cement, stone needed for building or maintenance) were in short supply.[9] One could not buy a cubic meter of wood in a cooperative store (*Konsum*) or lumberyard; rather, one needed to submit a request to the mayor for the material. He, in turn, went to one of the departments of the County Council, where he might well be told that there were only so many cubic meters of wood for the entire year to be divided among the communities of the county.

The process by which the county or the district distributed items in short supply among a myriad of requests was called "balancing," as was the process by which the mayor decided among the requests that came across his desk. Transactions that had been commercial in nature before the GDR now became bureaucratic, requiring mediation.

There were always long faces when Edgar returned to Göttern empty-handed, but Edgar found that the longer he remained in office, the better he got to know "his people" in Weimar and to judge whom one could best approach with this or that problem. The better acquainted with the system he became, the more possible it was to move the "balance" in the direction of Göttern.

The act of balancing epitomized Edgar's duties as mayor. The efforts he made on people's behalf in the face of a zero-sum situation mirrored those made by other officials further up the hierarchy. As Katherine Verderey points out, "Whereas the chief problem of economic actors in Western economies is to get profits from selling things, the chief problem for Socialism's actors was to procure things."[10]

Elections

Besides his weekly meetings in Weimar, Edgar met monthly with the Community Council. According to the constitution, the council was elected from the assembly. In compliance with the practice developed by the National Front, however, the first five to seven names on the unified list of candidates for the assembly, already predetermined by the front, constituted the council. In that way, there were no surprises. Each party and mass organization suggested eligible candidates. After scrutinizing them to determine whether each was a "good" citizen, the organizations forwarded them to the County Council for approval. From there, they went to the National Front. Because the county and district SED leadership was a member of the National Front, approval by the front carried with it the necessary approval of the SED.

Each eligible elector voted for this "unified" list on election day. The National Front's intention was to combine the particularities of the different parties so that the voters had an opportunity to vote for a uniform policy. The goal of an indivisible list was achieved by submitting to the voter names of persons who were essentially in agreement rather than in opposition.

At the polling station, each voter received the ballot and cast it openly into the ballot box. Although polling booths were available, no one was expected to use them, even though it was the constitutional right of every citizen to vote secretly.[11] Entering a booth often resulted in a check's being placed next to the voter's name. Entering the booth and crossing out a name or writing "not reliable" beside it led to problems for the voter, who would often be detained.[12] The election law of June 24, 1976, gave voters the right to request the removal of candidates from the voting proposal. Approval of the request was required from the National Council of the National Front.[13]

In short, the outcome of an election was predetermined before it occurred. There were no choices to be made between opposing candidates; only affirmative votes were possible. A dissenting vote could not be cast.

The chair of the election was appointed by the community assembly. It was his or her responsibility to see that eligible voters had a chance to cast a ballot. Persons whose names were on the voter list who failed to arrive at the polling station before a predetermined time were personally visited and requested to appear. A few chose to travel that day so that they would not be found at home. There were even those who openly boycotted the elections; these people were regarded as "unreliable citizens."

An election could be a stressful occasion for a mayor. He or she was ultimately held responsible for eligible voters who failed to vote from the unified list. If a person did not appear, the mayor would come under investigation by the SED. This happened to Edgar on more than one occasion and was the cause of considerable consternation. He was of the opinion that every person eligible to vote should fulfill his or her responsibility at the polls.

Edgar was satisfied with the selection of persons who, over the years, became members of the council but less so with the makeup of the assembly. He says, "The best representatives were members of the council. My relationship with the council was always a good one, and when you had a good council as mayor, you had only half the work. As for the twenty-three members of the assembly, half of them might as well have stayed at home. They either had no opinion or a bad one. Each party had membership in the assembly, so everybody was in there. We [the LPDP] had five, the CDU had two, but also the Women and the Youth were there. As for the SED, there were eight, and that's all there were in town. So they were all in the assembly whether they were any good or not. The SED was the leading party, and everybody knew that. You just couldn't get past Big Brother!"

The Budget

The council debated anything that pertained to the village, from the budget to the National Economic Plan. Its resolutions passed on to the assembly, and from the village to the County Council. As in every other area, the County Council had final authority in the matter of village finances.

As mayor, Edgar worked on the Göttern budget in concert with Frau Wiltrud Schmidt, a professional accountant in Magdala. She did the paperwork, and Edgar assumed responsibility for the outcome. Working up the budget each year involved making requests to the county finance committee: so much for the fire department, the day-care center, the sewage plant. In actuality, however, the county finance committee decided which items were to be funded and for how much. Once the funds had been allocated by the county, the community had some control over how they would be spent. A successful mayor saw to it that the allocations that became available every January 15 were spent by the following December 15. If not, the amount remaining would be subtracted from the next year's allocation. In such a planned economy, there was little room for maneuvering. The community could initiate only those projects that had no cost to the county—such as the building of a sidewalk for which materials and labor were being donated. Although communities were granted taxing powers, the returns from the real-estate tax, where low rates prevailed, and from certain fees, such as the dog tax, were insufficient to pay for significant projects. In such a constrictive funding situation, it took all the persuasive powers at Edgar's command to move his friends on the county finance committee in Weimar to lean in Göttern's favor.

The National Economic Plan

An important link between the community and the state apparatus was the National Economic Plan. Fulfilling the plan's obligations consumed a large share of administrative time in local governance.[14]

It is hard to overemphasize the role of economic planning in GDR thought and practice, being the main barometer by which the "triumph" of East German socialism was proclaimed yearly. Although each plan carried new expectations, the overall structure of the National Economic Plan as it affected Göttern changed little.

What the plan meant for Edgar as mayor was, first and foremost, a flood of paperwork. Directives flowing to him from the planning office in Weimar, establishing new goals for the coming year for each category of socialist productivity, were responded to with pledges designed to meet those goals. If the directive stated, for instance, that the 30 meters of paving of village streets laid the previous year was to be increased by an additional 35 meters in the course of the coming year, Edgar may have augmented the goal by another 5 meters, pledging the necessary labor power to get the work done. It was in Edgar's interest and that of his superiors up the hierarchy as far as Berlin to be able to state in print that the village of Göttern exceeded its street-paving goals by some amount. The value of the labor costs, approximately 5 DM per hour per worker, would be calculated as if paid by the town, seemingly Göttern's contribution to the plan. In fact, the work was "volunteered." Similarly, improvements in the kindergarten playground may also have been targeted as essential to socialist reconstruction. The rhetoric of the directive would point out that healthy young bodies and minds were a prerequisite for a productive socialist community and thus also an essential part of long-term economic planning. In this vein, the Weimar County Plan for 1969, the twentieth anniversary of the founding of the GDR, states, "The speed of societal progress depends principally on the training and educating of man towards a socialist personality."[15] Exempt from the play of market forces, economic planners constructed production goals based on an assumption of ongoing interdependence between all societal activities.

An economic form of planning that seemed like a shell game to orthodox Western economists was seen by GDR economists as a tribute to human intellect and the promise of human perfectibility.

Today, Edgar has little to say about the National Economic Plan, except to comment that it made exorbitant demands on his time until he realized that every plan was like the one before it, and all he had to do was change the figures.

Edgar's Support Groups

Although a mayor's effectiveness depended in good measure on how well he worked with the village council and the assembly, it was Edgar's opinion that without the support of the women, the youth, and the fire brigade, he could not have succeeded. It was not that he counted on these groups to support him at the polls—they would vote for him anyway, because they were obliged to vote and, by prior agreement, only his name would appear on the ballot as the mayoralty candidate. It was rather that Edgar needed members of these groups to carry out the numerous village projects that were an expression of his communitarian perspective, undertaken in the cause of civic betterment; "Cleaning Saturdays" were just one example of the community-wide tasks that Edgar organized.

Cleaning Saturdays became a hallmark of Edgar's mayorship, for which he chose to govern without resorting to politics. Politics in Edgar's mind meant coercively employing ideology. For him, the communal cleaning of roads and sidewalks, the raking of leaves, the clearing of brush could further village solidarity and village improvement without ideological harassment. It was not only the way Edgar liked to do things but also politically astute, "showing the others that we [in the LPDP] were better."

From time to time, an article appeared in a local or regional paper heralding Göttern's activities as evidence of an exemplary socialist community. Edgar relied on these reports to stimulate an even greater outpouring of energy for subsequent projects, especially on the part of local representatives of the mass organizations, the Democratic Women's Federation of Germany (*Demokratischer Frauenbund Deutschlands*, or DFD), and the Free German Youth (*Freie Deutsche Jugend*, or FDJ). As Edgar says of the DFD, "The strongest element in the village were the women.[16] They took home the things I was concerned about and discussed them around the dinner table. If the women said at home that this and that should be done, then that is what happened. That was a great advantage compared to other villages. Then you had to have the youth, too, but they were not so easy to mobilize."

The fire brigade was not a mass organization like the DFD and the FDJ. It came into being long before the GDR. It was—and still is—beloved by the people of Göttern, even though the scope of its activities has notably diminished since unification. Although not its primary function, the fire brigade, when called on, would fight fires—everyone in Göttern knows about the disastrous fire of 1833, its story having passed down from generation to generation. The major activity of the fire brigade was the sponsoring of parties, in conjunction with competitions between brigades in the laying out and reeling in of hose. The competition was intense. Doing well in a county or regional fire-brigade competition was a sure way to achieve local renown. As mayor, Edgar was active in fire-brigade activities. That the brigade had no ideological axe to grind endeared it to him. Edgar says, "After the competition, they would want to eat and drink in some remote *Gaststätte*, and I would always find the way to get money for that, even though it was not in the budget plan. I could only do this with the help of people who would give their signature, which meant that someone would sign up and be reimbursed for, say, fifty hours of volunteer labor, but only do twenty. The difference would be used for beer and bratwurst. Of course, the people in the Weimar County office were aware of this; they just didn't want to know about it officially. As mayor, I always had one foot in prison. [*Chuckles.*] Same for the young firemen; we had someone here in Göttern who was a professional fireman from Weimar, and he would train them for the competition in Bad Berka. But the fire brigade always wanted to have a lot of parties. They were very thirsty; 100 liters of beer was only the beginning. We would have a boar or piglets roasted on a spit, with two kegs of beer in the creek and ten bottles of Schnapps."

Informing the Citizenry

As mayor of Göttern, Edgar had the help of an administrative assistant for four hours a day as well as the use of a town messenger to go from house to house to let the occupants know of an approaching event and to ask for their assistance if any was required. Edgar says, "Today it all goes up on the bulletin board, and no one reads it. Before, people were better informed."

Edgar holds strong convictions about the importance of a well-informed and active citizenry. He says, "There were more opportunities for people to participate back then, and in the village, that was probably truer than in the cities, except where there was a good committee of the National Front. It was they who assumed the responsibility of motivating people. The idea was that the mayor would serve that same purpose in a small community—that is, to be in touch with people through close contact. Only people who are informed can draw their own logical conclusions about what is going on. That is just one of the things that has gotten worse."

Security

Although the law stated that the mayor oversaw the business of the local police, in fact, Göttern had no police force of its own. It did, though, have the services of a district patrol officer, who was shared among four towns. Edgar says, "Patrol Officer Werner was a nice guy, and because he was a nice guy, nothing happened. If an event was going on, he just came in, looked around, laughed a little, talked, and left. That is different today. Now there is a disco, and when somebody from Mellingen comes down, there is a fight and the police. The police won't be here. They only come when someone is half-dead or beaten to death, but not before that. We don't know where to go for help now."

As for the Stasi (*Staatssicherheit*), Edgar says he did not know who the informers were in Göttern but assumed there were some.[17] There were said to be between two and five in every town. If anything, being mayor exposed him more to their surveillance. He says, "I was in their black book anyway, because I had a sister in the West, and every month we had to report contacts with citizens from capitalist countries. There, I would write down when my sister came and how long she stayed. When my sister sent a wreath for the grave of my parents, it had to be reported. I knew them well in the cadre leadership, and they said, 'Oh, come on, you don't have to report that,' and I said, 'Yes, I do.' We also had to report when we received letters, but I didn't do that! Besides, my sister wrote to Wally, and that was better. But even so, it was all controlled, even the phone calls. I once mentioned that my sister had called me, and they said, 'We know that already.'

When my father retired, he would go west and spend four or sometimes six weeks, and they didn't say a word, either. 'That old man . . . ,' they would say. So nothing was eaten as hot as it was cooked [*Es wird nichts so heiß gegessen, wie es gekocht wird*]."

Political Training

Political training for mayors became institutionalized early on at the district and county levels. Every year, the Weimar County Council, in conjunction with the neighboring county of Apolda, organized a week of political training in Paulinzella, famous for its monastery ruins. In the summer, it was considered an ideal spot for mayors and their families to have a quiet and inexpensive vacation. During the training period, the heads of different county departments would make presentations, followed by discussion. The county leadership would present an evaluation of what had happened in the course of the year and inform the mayors about what to expect in the year ahead.

In addition to the annual county training in Paulinzella, the district organized month-long training every other year in Tambach-Dietharz. Edgar was permitted to go home on weekends, allowing him to take some of the burden off Wally's shoulders. In Edgar's opinion, a month of indoctrination "didn't do any good" and, in fact, may have had quite the opposite effect on participants.

Even more intense were the four-week training sessions at the so-called red cloisters in Frankenberg near Dresden. Edgar was required to attend twice in his capacity as director of culture in the Magdala community association. Edgar says, "It was really like a university, with workshops and lectures. We were only happy when someone came and talked to us about something practical. The political stuff was nonsense, and only those who were Mr. Big could talk [SED mayors]. There was nothing there for me; I could never understand it. They tried to teach us that in three to five years from then, we would all be communists. I said, 'I haven't ever seen a communist. To me, a communist is someone who does only good for the others and puts himself last. I will take my hat off to a real communist, because that [doing good for others] was what I was taught! What Marx and those guys had taught—that wasn't bad, but what they said was never put into practice.' Well, they just smiled."

Göttern in the Press

Looking at Göttern from the perspective of the local press during the latter years of Edgar's mayorship, one can gain some sense of the communitarian atmosphere that Edgar helped create and that the press underlined in its reporting. In the twenty-four articles about Göttern published in the local chronicle from 1972 to 1984, no mention is made of singular events, accidents, or natural catastrophes. Only socially affirming stories are printed—that is, accounts of men and women, young and old, working together to make Göttern a better place to live. In the local press, Göttern became a moral embodiment of what good intentions, hard work, and socialist zeal could accomplish against all odds—that is, shortages of funds and materials.

A news report dated May 20, 1977, recounts that the sixtieth anniversary of Red October was celebrated in Göttern as an occasion for sending personal gifts to "my best friend" in the Soviet Union. The 108th anniversary of Vladimir I. Lenin's birthday was commemorated by the women of the DFD and the youth of the FDJ, who were spring-cleaning Göttern's streets. In November 1974, a news release mentions that Mayor Edgar Schorcht was granted an award for excellence for his leadership in organizing the Join-In program. When asked to comment on the secret of his success, the mayor pays tribute to the National Front under the leadership of Comrade Rudolf Werner as well as to the women, the youth, the elderly, and the fire brigade for standing by his side.[18]

The Last Election

Edgar's last run for mayor took place in 1984, his intention being to designate his successor in 1987, when he reached age sixty-five. He would have been in office for twenty-two years. An article titled "Our Village in the 35th Year of the Republic, Pledge for Progress" comments on the election results as follows: "After all, Mayor Edgar Schorcht was sure from the beginning that it would turn out as foreseen: that the candidates for the Assembly presented by the parties and the mass organizations, and thoroughly reviewed by their work collectives (and ultimately by the National Front), would also pass the examination by the inhabitants on election day with flying colors."

One of the persons elected to the assembly was Edgar's daughter-in-law, Heidrun Schorcht. Heidrun had married Erhard Schorcht on January 28, 1972. Ten years later, she was granted her doctorate in American studies at the University of Jena. Unlike her father-in-law, she was a member of the SED and believed in its message.

Also in 1984, a time capsule that had been placed within the steeple of the local church, containing documents pertaining to Göttern's past, was taken down in the process of a complete restoration of the church's tower. The container included a report from 1864 as well as one from 1931. The latter was written by Walter Venus, Edgar's and Irmgard's teacher in the Göttern school in the 1930s. Venus, who became a member of the National

Socialist German Workers' Party (NSDAP) in 1933 and a party activist, writes about World War I: "What the German people had accomplished in [their] defensive fight against a world of enemies was superhuman." He speaks of the despair provoked by the inflation and the unemployment and ends by saying, "Our hope lies with our youth! They will tear apart the shameful treaty of Versailles and fight anew for a place in the world. Already the signs multiply that Germany will awaken! I will close my account with the wish that with the next opening of the tower ball the storm clouds will have moved on, and that the sun will again shine down on a free and happy Germany."[19]

At the opening of the capsule in 1984, a commission under the chairmanship of the mayor was appointed to draft a new "message" that would provide future generations with an understanding of the history of the village from 1931 to 1984. The commission consisted of Mayor Edgar Schorcht and his daughter-in-law Heidrun. The drafting of the message fell to Heidrun, who begins by noting that "the awakening of Germany which the chronicler of 1931 had hoped for did not lead this country to world recognition, but led the whole world into the Second World War, which cost the lives of 50 million people. The thunder clouds did not pass by, but concentrated into the night of fascism and started the darkest chapter in German history." She goes on to review how the major events of the past forty years had impinged on the county of Weimar: the arrival of the Americans in the village on April 11, 1945; the influx of resettlers from the East; the land reform and the splitting up of the state domain; the opening of the new polytechnical school in Magdala in 1952; the introduction of the first combine in 1955; the founding of the first LPG type II, named Gemeinschaft, in 1958, and the creation of an LPG type III in 1967; and the retirement of the last horse from service in 1968. As of 1984, 235 inhabitants lived in eighty-six dwellings and seventy-four apartments, 65 percent of which were equipped with interior toilets, 82 percent with bathrooms, and 24 percent with modern heating facilities.

The writer reminds the reader of the future and comments that 1984 marks the thirty-fifth anniversary of the founding of the GDR. She writes that "although we are quite satisfied with the increasing wealth and the security of the citizens of the country, we are at the same time worried about the tense situation in the world, which endangers all our achievements." She goes on to say, "Since the dropping of the first American atomic bomb in August 1945, a new, deadly danger has grown for this world. In striving for military dominance and the liquidation of socialism, the United States has developed ever newer and more dangerous weapons and has thus forced the socialist countries to increase military expenses. The large sums that are spent on weapons are thus lacking for the exploration of new energy resources, the fight against disease, and the protection of the environment. The nuclear weapons stockpiled on both sides can destroy the world many times over. A third world war would mean the end of mankind. Our only hope rests in the vigilance and power of reason of the people of this earth, and in the increased insight that we can all live in the future in peaceful co-existence."

The text of the message was read at the public town council meeting on August 29, 1984, and was hand-signed by both Edgar and Heidrun.

There is reason to believe that the final paragraph, confrontational in tone, does not reflect Edgar's sentiments. Whenever possible, Edgar opted for moderation and reconciliation. Even so, he could never free himself from the liens placed on him by the SED, to which the mayorship was beholden. On those occasions when the party pulled the strings, Edgar rose to the challenge in whatever way came to hand. But he saw those moments as capitulations.

Commentary

I first met Doris Weilandt, a devotee of art history, in the winter of 1993 at an illustrated lecture that she presented on Tuscan architecture at the Woman's Center in Weimar. Only later did I discover that she had also been the mayor of Göttern, following the term of office of Edgar's successor, Herr Reiner Scheidt. Knowing Doris to be a responsible person, I interviewed her in her home about the role of mayor itself as well as Edgar's performance in that office. Doris is the only person from Göttern, other than members of the Schorcht household, with whom I had a taped conversation.

Doris came to Weimar from Jena with her husband, a cabinetmaker, and their two young daughters. Tiring of city life, they chose to live in a small village. They bought their house on the *Anger* (village green) in 1983. Although an LDPD party member for one year, Doris is essentially an independent. She was voted a member of the village council in 1990 and became mayor in 1992.

Doris says of their coming to Göttern, "Edgar Schorcht was mayor, and I must say that Edgar was involved in helping us move here. The experience that we had with the mayor of another town was not so positive on the idea of someone from the city coming to his village to buy a house and live there. Edgar Schorcht readily agreed to go along with all of that. One needed a lot of permits and permissions, and he took care of it quite quickly—and spoke in favor of our coming here. We didn't have too much to do with the village administration at first, except in so far as it involved the reconstruction of our house. Regular windows were a problem. They were not obtainable through a craft shop; you would have to apply for your quota through the town administration, and you would be worked into a plan, and that might mean that in ten years you might get your windows [*laughs*]. Edgar helped us to get the windows.

"In many villages nothing happened at all, but Edgar just liked to get things moving. The maneuvering space for a community was really small; almost everything was prescribed. The agenda was usually fixed by the county authority. And when the community didn't get any money at all, then connections counted. Edgar, I must say, got a lot done. I really don't quite know how he managed to do so, but my house was all paid for by the county! There were no loans taken out, so Göttern now has no old debt."

Not everything, however, worked well for Edgar as mayor, as Doris points out. Although he was a good organizer, political tasks—such as assuming responsibility for elections—took their toll on him. Doris says of elections, "Everybody had to vote, and if you didn't go, you would get into trouble."

She speaks of one election in particular: "I know this from my own experience actually, because there was one election I didn't go to, and this caused Edgar a lot of distress. He almost got a heart attack. But I couldn't do it in any other way, nor could I explain to him that it was nothing personal I had against him. It was really bad for him. He had to carry the load.

"I know this from other towns as well; if you wanted your message [of not voting] to come across, then you would have to launch it before the elections. I did that, too. My message was passed on to the county. Some county secretaries came here to see my husband; I had gone away with the children, because the stress was just too much. They strongly urged him to tell them where I was, saying that I would have to go to the elections. And he said, 'But my wife won't go' [*laughs*]. They asked, 'Well, and so what is the reason for that?' I had talked to Edgar one more time [before I left]. He was paralyzed because of the fear; you know, it was tragic for anybody who held responsibility [like he did] when somebody didn't go to the elections. They had to write reports to I don't know where, and if they didn't get you, they would at least get the mayor for sure."

Doris tries to explain how difficult it was to get Edgar to understand why she could not vote. She says, "I am not sure what was going on inside of him. With this election thing, I was sorry [for him], but I couldn't do it, and he didn't understand it."

At the same time, Doris says, "Edgar was not an ideologue. But that does not mean that he was free of politics. Just thinking back to it, I would have had a lot of problems taking up just one of those [county] directives—not to speak of actually following up on it. I don't really know how he found his way through it."

According to Doris, a further source of contention for Edgar was the annual meeting known as the People's Forum, when representatives from the SED would report to the citizens of every village concerning the Great Line, the party's program for the coming year. As was the case with other "dissident mayors," the People's Forum was an occasion to diminish Edgar in the eyes of Göttern's citizens. About the People's Forum, Doris says, "I know there were a couple of things that must have been difficult for him. It was not Edgar who chaired the meeting. Rather, it was Rudolf Werner, head of the National Front, who was always fair to Edgar. Edgar told me of a couple of things, such as the petition to the state council [in Berlin] when the SED humiliated him in a nasty way, in a way that made him feel like a rotten nobody who had never done anything good for anyone. That was a situation in which the SED demonstrated that they were in charge and just tolerated him as mayor."

On the same subject, Heidrun remembers: "There was always a speaker [an agitator], a member of the SED, who presented the 'line.' Edgar always asked me to give a *Diskussionsbeitrag* on world politics, which I did to the best of my abilities, trying to make sense of it without repeating the *Neues Deutschland*, the SED paper. To someone from the 'other side,' my words would probably seem confrontational; I guess I saw the USA as the superpower in the offensive. Since then, I'm not sure which of the superpowers is the one to blame, but it always takes two to tango."

Doris extends her critique of Edgar in respect to SED ideology: "Anything concerning the Great Line—the blueprint of the state, which may not have been all that bad as an approach—was something that Edgar didn't even quite understand. He was more the social kind. He did the kinds of things that he enjoyed, retirees' parties or any kind of celebration where he could get singers or what have you from Weimar. That was his kind of fun, and he really grew with that. Then, too, he had his people at the county, whom he knew and got along well with. But with the Great Line, he didn't have too many things in common. He would say, 'That's how it is, and we'll just do it because there isn't anything we can do about it.'"

Nothing epitomizes Edgar's mayorship so completely for Doris as do his Cleaning Saturdays. She says, "There were a lot of Cleaning Saturdays, but they had as much of a social character as anything. People didn't have to go—they really wanted to do it. Edgar lived in the period—in this rural tradition where people didn't have TV sets—when village life was still happening under the linden trees, and when they still did things together and sat down over a beer at night. That was the impression I got."

Doris pays Edgar the compliment of attributing to him political skills that won him considerable community support as an "oppositional" mayor. At the same time, she takes him to task for not being able to stand up to those in command. The strength Edgar displayed as a popular LDPD mayor makes more disturbing, in her eyes, his pleading for her to vote regardless of her scruples, not so much out of regard for the difficulties she would face by abstaining but because of the criticism that would fall on him as a mayor who had lost control of his constituents. Part of the dilemma Edgar confronted undoubtedly inhered in the ambiguous role as an "exotic" mayor within the context of an essentially totalitarian polity. More of it, however, Doris suggests, lay within Edgar himself. The fact that Doris was able to freely commit a subversive act by not voting, one that provoked acute anxiety within Edgar, may reflect a generational difference between them in regard to deference to authority. An act that Doris saw as a defensible assertion of her "rights," Edgar regarded as an indefensible transgression of state authority. It may be that Doris is intuiting Edgar's essential Protestantism: his willingness to obey the state at whatever cost.

Despite the stress of the office, it is clear that Edgar's service as mayor of Göttern for twenty-two years was the high point of his adult life. There is no doubt that he savored the popularity he achieved and, most especially, the political success that his "nonpolitical" administration enjoyed in the face of SED intransigence. Despite his unwillingness to conform to party doctrine, he adhered to the principles of local governance, just as he dedicated himself to the advancement of the LPG once he assented to joining it. One can only conjecture about Edgar's life course if the Third Reich had not been defeated in 1945. What is worthy of our consideration, if only ironically, is the degree to which the GDR provided him with an arena for personal fulfillment.

14

From Stables to Apartments

Erhard, Heidrun, and Maria Move In

BY 1975, THE COURTYARD seemed deserted. The agricultural production cooperative (LPG) continued to use the barn for storage, but the large animals were gone. The cows had been moved to Magdala, and the bulls had been made obsolete by the artificial insemination station in Erfurt. With the death of Felix in 1962, the horses had gone the way of the bulls. The horse stall was temporarily made into a goose pen. The swine, geese, and hens continued to fill the yard with sound, but lacking in the cacophony was the striking of shod hoof on stone paving, neighing, and mooing. Edgar and Wally were relieved to be free of the arduous work involved in the feeding and caring of the horses, bulls, and cows, but the excitement created by their presence, the very sound of them, was missed.

Like their neighbors on the *Anger*, the Schorchts were confronted with an excess of empty space around the courtyard, something they had never expected to experience. Heretofore the concern had been where to fit in the additional newborn stock, the new machines. No purpose would be served now in leaving the structures idle, only to have them deteriorate through abandonment. But to what end to put them was not clear.

It had been obvious to Edgar for some time that the East was never going to catch up to the West, despite declarations to the contrary emanating from Berlin. The gap between the two Germanys only reaffirmed the impression of Soviet backwardness that Edgar had gathered while serving on the Eastern Front. At the same time, it was certain that there would be no return to private farming, whereby Edgar could regain control of his holdings, equipment, and livestock to manage as he saw fit. To all intents and purposes, the Schorcht farm had ceased to exist.

What remained under Edgar's control was the farmstead, the house, the courtyard, and the encircling barn and stables. Most of Edgar's daily responsibilities lay with village administration, not with farming. Wally continued to work with the cows, but in the collective stables in Magdala. Their older son, Erhard, wanted nothing to do with agriculture. His interest lay in technology and photography. As soon as he finished high school, he headed toward Jena and a training program, sponsored by Zeiss Optics; later, he enrolled at the University of Jena. It was the childhood dream of their second son, Norbert, to follow in his father's footsteps, but upon completion of his training in agricultural engineering, he was assigned to a position in a factory for agricultural machinery on the outskirts of Erfurt.

Only Roswitha continued to live at home while she trained as a seamstress in a cooperative workshop in Weimar. The house, like the stables, was emptying out. The stem family cycle, which would have seen Norbert if not Erhard join with his father to work the farm and eventually assume control of it, was no longer operative. The future was uncertain. Still, Edgar's penchant for transforming an unlikely situation to his and to others' advantage, as he had so often done as mayor, was as focused as ever.

It became engaged when he learned that his son Erhard and his daughter-in-law Heidrun could no longer continue to live in a single room in a University of Jena dormitory with a baby on the way. What better solution for everyone could there be, Edgar thought, than for the young family to move in with Wally and him? Even if Erhard would not be able to inherit the farm, he could live on the farmstead with his wife and child, as the oldest Schorcht sons had always done.

Edgar and Wally had learned, however, that you could not be too insistent with the young; if you were, they balked. He and Wally were pleased when Erhard and Heidrun accepted the offer. Unable to find lodging large enough for their needs, they had no place to turn. It was agreed that they would move into the first floor. No major renovations would need to be made other than building a bathroom at the end of the hallway, where a door led into a hayloft above what had been a stall. The full bathroom that Edwin had installed years ago was on the ground floor behind the kitchen. With a baby due, Heidrun wanted one closer to hand.

The new bathroom was built in the former hayloft, while across from it a darkroom was made for Erhard. Wally's mother, Frieda Hünniger, vacated the large bedroom at the front of the house so that it could become the young people's living room. It needed only shelves to hold all their books. The small kitchen installed for Frau Hünniger, which she had never used, now became theirs. Erhard and Heidrun would sleep in what had been Edwin's room.

Later, when the baby grew up, they would make the storage room into a child's room. Meanwhile, the baby would sleep with them. Edgar and Wally would take over what had been the extra room for their own bedroom; otherwise, they would live on the ground floor, leaving the first floor for the young people. Erhard and Heidrun moved into the house in November 1977. A baby girl, Maria, was born on January 4, 1978.

The Lowering of the Barn

For some time, it had been decided that the height of the barn posed a hazard and needed to be lowered. A magnificent structure, one of the tallest in Göttern, it helped make up in storage space that the cramped courtyard lacked. Half-timber in construction, it was four stories high. Hugo, Edgar's grandfather, had the barn built in 1913, employing Paul Hempel, a carpenter from Magdala, to do the construction.

The reconstruction started in 1984. Edgar, reelected as mayor for his last term in office, arranged to put aside some time to work on the barn. Wally, retired in 1981, was now at liberty to help, although she was somewhat hindered by her infirmities. Erhard, Heidrun, and Roswitha helped whenever they had the opportunity. Norbert came over from the village of Hetschburg, 10 kilometers southwest of Göttern, to lend a hand. Indebted to his father for the assistance he had given him in subsidizing the cost of his house in Hetschburg, Norbert was more than willing to help out his father in return.

The first task in the reconstruction of the barn was to disassemble the half-timbered construction by pulling loose the stones and bricks that filled in the spaces between the wooden supports. It was arduous and dangerous work on the high ladders. Hands got scraped and bloodied. Edgar and Wally claimed that their thumbs remained numb for several months afterward.

The most precarious task was removing the beams one by one. For this, Edgar called in his friend Alfred Rath, a carpenter from Magdala, and his two sons. Each beam had to be correctly positioned by block and tackle for dropping. When all were down, Norbert, the best at math, calculated that 22 cubic meters of solid wood had been removed.

Then began the job of reconstructing the new lower roof. Wherever possible, the old wood was reused. The crossbeams were put in place and tested, followed by the rest. It was hot work, and the unusually warm summer of 1984 did not help matters. When the new roof was in place, a pine tree was fastened to the peak of the roof for the traditional "topping-off," to the applause of all below. Then the beer was broken out, and as much wurst as anyone could eat was roasted. It was the first time that the family had worked together as adults. Amid the laughter and the joking, there was talk about what project they could take on next.

Before the year was out, the old chicken coop was converted. Half of it was made into a storage room for preserves, the other half into a garage for Edgar's Trabant. Now that Wally was retired, she had time to put up plums, cherries, blackberries, and raspberries, and with so much room to store the jars, she could preserve to her heart's content. Edgar had acquired the Trabant in 1985 after waiting twelve years for it. It was a welcome change to his old DKW, which while reliable was never as comfortable as the Trabi. Nothing seemed to be so much a harbinger of the future as was the building of that first garage in the courtyard.

An Apartment for Roswitha

In February 1978, at age nineteen, Roswitha gave birth to a daughter, Katherina. She had met her husband at a dance three years earlier, and they were married in June 1978. For the first two years, the marriage went well. Then problems began that led to their divorce in 1983.

Roswitha, fleeing a broken marriage, moved back to the security of her childhood home. A year later, Roswitha met Manfred Netz at a harvest dance in Bucha. They found that they had much in common. Manfred had also been married before and had a daughter by his first wife. When they felt comfortable with each other, Manfred moved in with Roswitha and Katherina. Roswitha was living across the courtyard over the washroom, where the "old men," Louis and Hugo, had once lived. Roswitha and Manfred were married in October 1985 in the Weimar City Hall.

Everyone agreed that those old rooms were not fit for a young family to live in. At first there was talk about doing them over, but Roswitha disagreed. There were too many associations with that space. Not only had she lived there with her first husband for a while; so had Erhard and Heidrun, and Norbert and Renate had spent time there before they were married. When the rooms were not occupied, they were used for apple storage during the winter months. Roswitha wanted to make a fresh start, and she proposed constructing an apartment in part of the barn that Edwin had built across the way in 1939.

It was Edgar who decided what would be done for Roswitha. Not only the apartment above but also the washroom and the horse stall below would be rebuilt into a larger apartment for Roswitha, Manfred, Katherina, and the baby that Roswitha was then carrying. Farther along the courtyard, where the stall for the young stock had been, there would be space for another garage, maybe two. It immediately seemed to everyone like the appropriate thing to do.

Edgar would get an architect from Bad Berka, who he was certain would not charge too much. What was going to be difficult was getting hold of the supplies. It was fortunate that Edgar was mayor and knew his way around. All the same, he could not show favoritism and give the Socialist Unity Party of Germany (SED) ammunition with which to snipe at him.

Before the final blueprint was drawn up, it seemed as though everyone in the family had had a hand in the planning. There was general agreement, though, that before they were through, almost everything would have to be rebuilt, starting with the washroom. It took a while just to remove the ovens, the milk cooler, the big boiler, and the cauldrons used for hot water in the slaughtering. When everything was torn out, the old floors were blasted with a pneumatic hammer, and the stairs were removed. Roswitha, well along in her pregnancy, had to negotiate a ladder to get to the bedroom. New girders, supporting beams, and sills were installed as the material became available. There were times when the work came to a complete standstill; those were the moments when Roswitha wondered whether it would ever get done.

On February 9, 1986, a baby boy, Philipp, was born to Roswitha in the Sophienhaus in Weimar. It was Edgar and Wally's fourth grandchild and first grandson. Roswitha stopped working at Zeiss in Jena, where Manfred also had a job, and took her state-allotted maternity year.

The apartment, consisting of living room, three bedrooms, study, kitchen, bathroom, and terrace balcony, was completed just before Christmas 1987, a year and a half after it was started. Work continued for several more months to convert the last remaining stall space, the former pigsty, into a family room, where Edgar, Wally, and their children and their families could meet to observe birthdays, *Jugendweihe* (youth consecration), *Kirmes*, and other holidays.

Christmas that year was celebrated in Edgar and Wally's living room. There was just enough room to hold all the Schorchts, including Norbert and Renate and their two daughters, Magdalena and Julia. Renate was pregnant with their third child. They drove over in the morning from Hetschburg in their Trabant, which Edgar and Wally had helped Norbert finance.

The children—Katherina, Maria, Magdalena, and Julia—took their places around their own small table to eat. Philipp sat on his mother's lap. Edgar, not given much to piety, did say how fortunate he and "Mutti" were to have so many beautiful grandchildren to play with now that they were both retired. He added, too, how comforted they felt knowing that their children were settled with careers and jobs and places to live to raise their children.

Indeed, everything worked out differently than anyone had ever thought it would. At the time of Erhard's birth in 1951, it had not occurred to Edgar and Wally that a son of theirs might choose to not take over the farm. Even more remote from their minds was the chance that a day would come when no one in the family would be farming, while the courtyard surrounded by living quarters and parking spaces for cars would be the center of an extended family that they themselves had helped bring about. Equally unlikely was the possibility that the Schorchts' fields would be cultivated by a private enterprise.

The New Agrarian Cooperative

When the inclination moves him, Edgar drives to nearby Niedersynderstedt to visit with Herr Reinhard Schüffler, the young general manager of the agrarian cooperative *Agrargenossenschaft Niedersynderstedt* (AGN). As an associate of AGN, Edgar receives a share of the returns from produce sold.

On the outside, the stables, pigsties, fodder-storage barns, farm equipment, and sheds look the same as they did when Niedersynderstedt was the headquarters of the Cooperative Department of Crop Production (KAP) in the last years of the German Democratic Republic (GDR). Now, one-eighth of the personnel who manned it when it was party-controlled is sufficient to operate it as a private cooperative. The cooperative's large size and productive capacity allow it to compete favorably with smaller independent farmers in the West. Herr Schüffler takes pride in knowing that AGN products now circulate not only through outlets in the rest of Germany but through European Union (EU) markets in western Europe as well. Edgar shares Herr Schüffler's gratification yet feels frustration at the lost opportunities for profiting from the land. He says, "You know, if we could have sold the land after the war and put the money into the bank, I would be a rich man now. But instead, it all ended up in the collective. We worked very hard then. We did well and produced a lot, but now we don't have anything to show for it. That is something the people in the West don't understand. They worked for themselves, they got money, they got rich, and now they come over here and try to tell us that we didn't know how to work. My sister in the West is the same way. She never really understood us here. Then, too, even if I had wanted to keep us going as a farm after '89 for Erhard's or Norbert's sake, I couldn't have. It takes at least two to three million [deutsche] marks to start farming again, and 30 hectares are not enough to get you anywhere. Now you need at least 100." After a moment's reflection, Edgar says, "It's nice, though, for an old farmer like me to see those fine big fields being farmed out there by the cooperative. I know how hard it was to make them in the first place."

For the most part, Edgar stays close to home, being near Wally in the kitchen or sitting with her in front of the television when he is not out puttering in the barnyard or working in the garden. Edgar's main interest now lies in the household that he and Wally helped bring into existence.

Commentary

When I first began to think about studying an East German family, I did not imagine that the family I would find to work with would be an extended one. To my mind, such arrangements were associated with familistic traditions of southern Europe, not with the individualized aspirations of the industrialized North. And yet having written about an extended household in the village of Sermoneta in southern Italy in 1977–1978, the prospect of working with an East German extended family was compelling. What was at stake was further understanding of the place of multiple-family households in modern European societies.

In my previous study of the Italian family, *The House that Giacomo Built*, I recall the surprise I felt upon returning to Italy in 1977 and finding Giacomo Savo putting the finishing touches on a house he was building nearby for his eldest son, Eleuterio—soon to be married—and the plans he had for building another, nested above his own, for his second son, Massimo, against the day when he, too, would marry. I was surprised because the formation of an extended household seemed to be an anomalous response to the surge of economic prosperity that had begun to materialize in Italy in the 1960s. What Giacomo and Maria were doing appeared to fly in the face of the received wisdom that holds that affluence selects for small family structures, not large familial arrangements. The Savos were not alone. All through the countryside surrounding Sermoneta, parents like Giacomo and Maria were providing houses for their sons preparing to marry. The reason soon became clear: it was not considered fitting that a newly married couple should start their married life living here and there like wandering nomads, throwing good money after bad for rent.

That was in 1977, the year Erhard and Heidrun moved in with Edgar and Wally in Göttern, to be followed by the birth of Maria in January 1978. When Manfred joined Roswitha and Katherina in 1984 in the "old men's" apartment across the courtyard and then married in 1985, the extended household was complete. Along the *Anger*, other families had converted former stables into apartments. By 1984, fifteen such dwellings had been constructed in Göttern for young couples. The rationale was clear: the chronic lack of housing in the GDR made the conversion of unused stables an appropriate response to a son's or daughter's need for a place to live when he or she married. What was ironic, however, was the emergence of a traditional household as a solution to the dislocations caused by the GDR project to "modernize" the countryside through industrialization.

Despite the convergence of form in the Savo and Schorcht households, there are, as one might imagine, important differences in regard to organization and meaning. Both Maria's and Giacomo's progenitors were agricultural day laborers (*Bracciante*), for whom there was no patrimony to be transmitted to their offspring. Afflicted as they were by poverty, each of their children upon marrying was constrained to leave the parental home and settle him- or herself as best he or she could. It was only after Giacomo won the ownership of 3 hectares of formerly communal land in a lottery and built a family home there that he was able, through the sale of one of these precious 3 hectares, to realize an ideal by building houses nearby for his sons. Giacomo and Maria's two daughters would in time be accommodated in houses built by their respective fathers-in-law, proximate to their own.

Now the Savos, living out their lives in one another's company, give meaning to the mutuality of their existence by means of a familistic rhetoric. Much is said about their being a united family. Striking out on one's own, becoming independent, is not a value to be pursued. On August 15 of each year, the Feast of the Assumption is marked by the return of Maria's sister Teresa and her family and her brother Giuseppe and his family from the North to celebrate their kinship. The extended household structure that Giacomo and Maria have devoted their married lives to constructing is the fulfillment of a familistic ideal made possible by a newfound prosperity.

Since their arrival in Göttern, the Schorchts have known toil but not want. Blessed by a patrimony of land, they cyclically experienced a version of an extended household when the inheriting son (Johann Wilhelm, Hugo, Edwin, and Edgar) was joined by his respective wife (Adelheid, Alma, Elly, and Wally) in the parental home at *Anger* no. 7. The enlarged household was informed not by familial sentiment but by the legal language of a contractual relationship specifying the transmission of the farm from father to son in return for his support of his parents until their deaths. Then, with the full collectivization of agriculture in the 1970s and the demise of private farming, Edgar and Wally, freed from the exclusive requirements of impartible inheritance, were able to provide for Roswitha what Edwin was not allowed to allocate to Irmgard: an equal share in the inheritance. In this respect, the organization of the new Schorcht household differs from the equivalent in Italy, where patrilocal residence allows sons to marry in and requires daughters to marry out, allowing the men to remain proximate to the ultimate source of nourishment.

Edgar and Wally's invitation to Erhard and Heidrun to live with them found acceptance in its appeal to mutual convenience. It was in the same spirit of instrumentality that an apartment was constructed on the other side of the courtyard for Roswitha, Manfred, Katherina, and Philipp, providing with living space an acceptable solution to the problem of unused space.

The creation of an extended family by the Savos in the 1970s and another by the Schorchts in the 1980s suggests the versatility of that structure in meeting needs posed by life in contemporary society, whether those driven by capitalist market economies or by socialist planned economies. To put it somewhat differently, a comparison of the Italian and German examples provides reasons to believe that household forms may assume a greater diversity of structures than modernization theory generally allows.[1]

PART III

15

Erhard and Heidrun Schorcht

Erhard: The Early Years

ERHARD SCHORCHT was born in the hospital in Jena on June 24, 1951. His earliest memories have to do with the big kitchen at the family homestead in Göttern, where his mother, his grandmother, and the female refugee from the Sudetenland, who cooked fragrant Bohemian dishes, held sway. He remembers seeing his grandfather Edwin at the head of the table and marveling at how big his hands and feet were. Grandfather Edwin seemed like a quiet, serious man, not given to bantering. When he did talk, it was almost always about work, and how there was nothing more important for a man either to do or to think about. When Erhard told him about adventures he had had, his grandfather would say, "You'd do better to study for school," and when older, making a reference to women, Grandfather Edwin commented, "Women are a chain on your leg when it comes to a career."

Erhard's other grandfather, Erich, sometimes sat at the table, too, where from time to time he would burst into laughter for no apparent reason. Erhard recalls that if the man did not laugh, everyone thought something was wrong. When he would come to visit, he often brought something new to show Erhard. He was the first whom anyone knew to have a tape recorder and, later, a television set. His wife, Frieda, Erhard's maternal grandmother, "was cut from different cloth." Erhard recalls her as frugal and industrious and says she worried a good deal about whether they would be able to make ends meet.

Erhard's paternal grandmother, Elly, played a special role in the memories of his childhood. He recalls her as always working hard and yet looking at the bright side of things. "We understood each other well, my grandmother and I," he says. "It may be because I was the oldest, but she always seemed to have a special place in her heart for me. Of course, there were some things she didn't like, such as when a bunch of us kids wanted to play in the house. She would tell us to go outside and have fun. She sometimes told a little white lie to get us to go out, like, 'Up there on the hill a glider plane just landed; go and see it.'"

Erhard can no longer remember the names of all the persons who frequented the kitchen at mealtimes, especially at harvests. There were always extra helpers in addition to the regular hired hands. One man stands out in his mind because he apparently had something to do with the events of June 17, 1953—the East German uprising. Erhard was too young at the time to remember what was being talked about, but he recalls stories of how the Stasi (*Staatssicherheit*) went to this man's apartment in Jena and examined his radio to see which station he had been tuned to. Erhard says, "I didn't know what to make of those stories at the time."

Erhard's Siblings

In Erhard's view, Göttern was a good place to grow up, with many places to play: in the woods, on the hill, in the fields, along the banks of the Magdel, or in someone's barn. Whatever was forbidden was the most fun. Sometimes he would disappear for the whole day and come back covered with dirt from head to toe. Erhard says, "My mother was ashamed of me because I was so dirty. One of my friends from the neighborhood was always clean, and that made it worse."

Erhard was four when his younger brother, Norbert, was born. Of that event he can remember nothing: "At one point he was just there." For Erhard, Norbert was more a source of pain than pleasure. When Norbert cried, the blame fell on Erhard. It got so that when Erhard wanted to hit him, the younger brother would cry in anticipation of the blows raining down on him, causing Edgar to go after Erhard anyway. Norbert was just too young, in Erhard's opinion, to join in all the activities that he and his friends liked to do. The fact was that more children were born in Göttern between 1950 and 1952 than in subsequent years, leaving Norbert bereft of age-mates.

With the birth of his baby sister, Roswitha, on August 13, 1959, it was different. The eight-year distance between them offered less reason for strife. He expressed his major concern about her when he asked his mother whether girls were born with legs, too, as boys were. Erhard's worry had to do with Roswitha's potential aptitude for soccer.

School

When he was three, Erhard went to kindergarten. It was a blessing in a way for his mother, for the demands made on her time by farm work left little time for her to spend with her son. The kindergarten in Göttern, the so-called Harvest Kindergarten, was begun in 1954, when the elementary school was moved out of the room where it had been held in Erhard's grandfather's and father's time. Its sessions were scheduled to correspond to the seasons of the year, when farm work was at its heaviest and mothers, kept busy in the stable or field, had little or no time for their children.

In the early 1950s, the one-room schools that had existed from the mid–nineteenth century in the eight villages constituting the Magdala Community Association were consolidated into one large elementary school in Magdala. No longer would children of all ages be taught by one teacher in one room. In its place, a new school was completed in 1952, providing space for seventeen classrooms and ninety students.[1] Gone, too, were the pedagogic titans, who knew every subject and were versatile in the playing of a variety of musical instruments.

The German Democratic Republic (GDR) regime, still in its infancy when Erhard began school, was determined to make a clean break with the educative practices of the past. The new policy, put forth in the "Democratic School Reform" of June 1946, proclaimed the inauguration of a single-school system designed to provide educational opportunities for children from all social classes and to develop in them "deep love for their socialist fatherland, respect for people, love of work, and class solidarity."[2] At the center of the reform was the ten-class general polytechnic secondary school, which had as its mission the education of the socialist citizen through the proper integration of theory and practice in the sciences and the humanities. The subjects taught were prescribed by the Ministry of Education. In the first two grades, the children focused solely on German and mathematics. In the third grade, German grammar was introduced, as was local geography.

Erhard remembers being taught sewing in the third grade by a woman from Göttern, whom he still passes in the street from time to time. By the fifth grade, classes in Russian, history, biology, and geography began, and in the sixth, physics was introduced. The polytechnical side of the education—the applied classes—started in the seventh grade. Erhard learned how to work with iron—how to saw, file, and shape it—and how to operate a variety of machines. He also took a course in applied economics, called Introduction to Socialist Production, in which students were taught "basic business principles." In fact, Erhard says, "we were actually taught very little."

Like all the other children entering school, Erhard became first a Young Pioneer (*Jungpionier*) and then an Ernst Thälmann Pioneer (*Thälmannpionier*), remaining one until the sixth grade.[3] Because membership was taken for granted, most children, in Erhard's opinion, had little idea of what it was all about. On Pioneer assembly days, every child wore his or her Pioneer kerchief and repeated the Thälmann oath: "Ernst Thälmann is my shining example. I swear to learn, work, and fight as Ernst Thälmann taught. I wish to behave according to the Thälmann laws, faithful to our greeting. I am always prepared for peace and socialism." On these days, they often listened to an exhortative speaker. For the most part, extracurricular Pioneer activities consisted of collecting money for the impoverished in developing countries allied with the GDR.

Russian, which began in the fifth grade, continued until the tenth. Erhard resented having to take it, not because he found it difficult but because it was obligatory. Even though it was a required course, hardly anyone spoke the language. Erhard remembers a joke that made the rounds: "What does a miniature poodle from Bavaria and a graduate of the Polytechnical High School have in common? Neither of them can speak Russian!" Furthermore, unlike English, the language was tainted politically. It was the language of Big Brother.

The teachings of Karl Marx and Vladimir Lenin were presented in a civics course taught in grades seven through ten—"a beloved subject," remarks Erhard with a smile. As he recalls, "It was not unduly pushed upon us," but he admits to not being able to remember much about it other than that it was highly philosophical and that the teacher had a nervous tic.

The Way Home from School

Taking the bus was the quickest way to get to school in Magdala. When for one reason or another the bus did not come, the children would walk, with Erhard supposedly keeping an eye on his younger brother. The children hardly ever took the bus home because of the pleasure in meandering along the way: running in the fields, jumping across or sometimes even falling into the stream. Springtime was the best, when the water was running high and the air was full of sweet smells and birds. On a lazy day in June, it might take the Schorcht children the whole day to get home, ensuring that there would be fewer chores left to do when they arrived but guaranteeing that reprimands would be heaped on them.

Erhard's first farm chore was cleaning the milk cans. The milk destined for Jena was picked up early. After the empty cans were dropped off, it was the job of the youngest child in the family to retrieve them and wrestle them onto a cart. There were usually seven cans, but they did not all fit that readily, so it was sometimes necessary to load them on the run after the cart had started. From loading milk cans, Erhard graduated to cleaning the stalls, a task he did not enjoy.

Joining the Free German Youth

On leaving the ranks of the Pioneers at the end of the seventh grade, all students were expected to join the Free German Youth (FDJ) by age fourteen.[4] At first Erhard chose not to join, adamantly disagreeing with the teacher's warning that everyone had to become FDJ members upon entering eighth grade. "No, no, we don't have to," rejoined Erhard, taking satisfaction in thumbing his nose at the system. It seemed that if you did not always do as you were supposed to, nothing happened. Erhard came to the realization that, as his father often said, "There are lots of things that you don't eat as hot as they are cooked." Erhard was well aware, though, that by the ninth grade all students had to apply for career training, for which FDJ membership was required. Thus, he became a member.

Erhard found that belonging to the FDJ made demands on him far in excess of anything he had experienced as a Pioneer. Although originally a nonpartisan organization expressing "antifascist unity," by the time of the Stalinization of the GDR, the FDJ had become the training-and-recruiting ground for the SED.[5] The level of indoctrination was most strident during the "FDJ School Year," a supplementary course of political training that met periodically during the semesters, a kind of "ideological Sunday School," as Erhard calls it.

Erhard regarded the political education that he was subject to as similar to religious education, in that both, even though serious in intent, were essentially irrelevant to the everyday concerns of the students: "We avoided it as much as we could, but you couldn't always avoid it, and then we just participated . . . but it was artificial and spoke to hardly anyone." Erhard likens the way they felt about what they were told to the way they felt about their blue shirts: "We had blue shirts that we were supposed to wear, but we weren't motivated to wear these blue shirts. When you didn't have to wear it for a public event, you ruffled it up in a tight ball. You didn't wear it to proclaim your pride but because you had to. There was none of the kind of conviction, the ideal attachment to country, that we knew from American films."

Erhard draws a parallel between his resistance to the FDJ and his feelings about the German-Soviet Friendship Union, which he joined later in the university: "It was little more than a coffee club. It wasn't very important. It promoted friendship with certain people, but it stifled it with others because you were forced to be a member. It is always like that when you are forced to do something. I would have enjoyed a German American or German French or German Arabic connection. The disadvantage of the whole thing was that it was prescribed. As for America, our impression of it came from *Bonanza*, from TV, from the movies. There were more American films in the theaters than anything else. Personal experience with America was almost unheard of. Maybe a few books, hearsay, that was all."

Erhard remembers the authorities initially saying, "We have arguments; we can prove it." Later they would say, "That is not our method—here are the facts"; eventually they would just say, "Believe us." Erhard says, "It seemed that the more they wanted you to believe, the more ceremonies and parades they had."

Of being in the system, Erhard says he took much of it for granted, comparing his experience to taking a bath: "When I'm in the bathtub, little by little I let in the hot water. Eventually it gets quite hot, but I can stand it, whereas if I were to fill the tub and hop right in, I would be scalded." Erhard remembers that the further he went along in his education and training, the more important it became whether one was in or out of the FDJ. Indeed, he began to look back on his earlier, more cavalier attitude with some amazement.

In the ninth grade, Erhard took the admission test in Jena for entrance to a Zeiss program, designed to qualify students for their *Abitur* (higher education entrance exam) as well as train them for a career. It was, he remembers, a day in November. He felt the cold as he rode to Jena on his motorbike.

Work on the Farm

By the time he was twelve, Erhard was already large for his age. He was proud of his ability to carry feed bags weighing 30 to 40 kilograms. During harvest time, there were potatoes to be dug by hand and beets to be thinned out, tasks no one liked to do. Haying was an especially busy time: loading and unloading the hayrick, trying to beat the rain. Before Erhard was old enough to rake and fork the hay, his father would allow him to steer the tractor in low gear while Edgar pitched the hay onto the wagon.

Erhard's last experience with agriculture occurred in 1968, the year he finished the tenth and final grade of the polytechnical school in Magdala. He was then ready to begin career training. Although he had not yet decided on a specific career, he was certain it would be not agricultural but most likely something technical.

In the last two years at the Magdala school, Erhard and his friends began to effect the sixties look. It was mostly through classmates and Western television that they were able to pick up on pop culture. A class picture taken on the steps of the school in Magdala shows Erhard with his grown-out hair. In his opinion, it was not too long, but there were comments in the family, especially when Norbert began copying his older brother. Erhard also remembers taking a pair of his father's pants to widen the legs of his own to give them the bellbottom look. *"Das fetzt!"* ("It's mindblowing!") was the jargon used to express approval of such successful fashion initiatives.

Zeiss Career School

In the fall of 1966, Erhard was admitted to the Zeiss Career School in Jena, a program in precision mechanics that would grant him the *Abitur* necessary for admittance to the university and train him for a specific job with Zeiss. Actually, the curriculum was weighted on the side of job training. There was an initial month of schooling in which art and biology were cut back to make room for additional training. Biology and drawing were finished early, and there was no music at all. Replacing them were courses in technical drawing, business courses, and metal-working labs.

The initial month was followed by another totally devoted to training in pre-mechanics, which consisted of practice in milling, turning, drilling, cutting, and polishing. The goal was flexibility, so that the student was prepared to switch from one operation to another. By the third year, the training was carried on in a factory, where the students, in addition to acquiring skills in mechanics, were trained to understand the workings of the factory—everything from maintenance to finance.

Erhard Meets Heidrun

It was August 1967 when Erhard, seventeen at the time, first met his classmate Heidrun Matthies, then sixteen. Erhard admits to having had "many other acquaintances," going back to the ninth grade in Magdala: "When we could ride motorcycles, that's when it started. It didn't work if you arrived on a moped and it only had one seat." Erhard recalls looking around the classroom that first day of school in Jena: "My first thought when I saw the class was "Hmm, not too much going on here." It was not long, though, before Heidrun, with her forthright manner, blonde hair, and blue eyes, attracted Erhard's attention.

Heidrun

Heidrun Matthies, born on April 27, 1952, came from a farming family in Langenhain, some 80 kilometers to the west of Göttern. Heidrun's mother's family was from Silesia. They all left before the war, with some of Heidrun's aunts going west in search of husbands, while her mother followed an older sister to Gotha. They were later joined there by Heidrun's maternal grandmother and grandfather, who was a bricklayer.

Heidrun's father's family had been farmers on the same land in Langenhain since the end of the Thirty Years' War. "We were not poor peasants; we always had enough to eat," is the way Heidrun describes the family situation when she was growing up. When not in school, Heidrun and her younger sister worked with their parents in the stalls and in the fields. Her father raised wheat, rye, oats, potatoes, turnips, and pigs on his 10 hectares and sold the milk from his cows. The plowing and hauling were done with the help of a pair of oxen. One of the family pictures that Heidrun has in her possession shows her and her parents sitting in the middle of a potato field, eating in the company of helpful neighbors.

In the 1930s, Heidrun's father and uncle joined the *Sturmabteilung* (SA) and the *Schutzstaffel* (SS), respectively, much to the dismay of their mother. Heidrun's grandmother worked for a Jewish business family in Waltershausen, whom she held in high regard. She was an outspoken woman and took strong offense at her sons' castigation of the Jews. She was warned to keep quiet or she would get into trouble. This was recounted to Heidrun by her father. Heidrun says, "In a small place like Waltershausen or Langenhain, you just belonged to the SA or the SS. It wasn't so much a matter of conviction; you went with the group."

Heidrun's father was held prisoner by the Russians for several years after the war. He returned to Langenhain in 1950, broken in health, having worked in a mine. His brother failed to come back from Stalingrad. While convalescing at home, Herr Matthies was taken care of by a nurse. One year later, he and the nurse—who would be Heidrun's mother—married.

For several years, the couple struggled to meet production quotas. When they joined the agricultural production cooperative (LPG) in the late 1950s, they found they preferred working with others and were relieved, as well, to be freed from the burden of taking care of the animals. Excused from fieldwork due to his precarious health, Heidrun's father became a tractor driver, adept at maneuvering heavy loads in difficult terrain.

When growing up, Heidrun felt especially close to her paternal grandmother, Frieda. She says, "In peasant families, no one shows feelings; they are reserved. But my grandmother always took a special interest in me because I wished to become a teacher. I went into teaching partly because of her. She used to tell me that she was proud of me."

Heidrun remembers with warmth the conspiratorial pleasure her grandmother took in sheltering her and her sister in her room when their parents had neighbors for an evening, all three peeking down at the company through a hole in the floor. Heidrun admits to being overwhelmed with emotion for the first time in her life when, at twenty-five, she cried audibly at her grandmother's funeral.

A picture in Heidrun's possession shows her at age six, on her first day at the Langenhain polytechnical school, holding the traditional *Zuckertüte*—the paper cone, almost as large as her, full of goodies given to her by her parents.

Heidrun has another picture of herself at age fourteen in her black confirmation dress, her hair in a long braid down her back. "Protestants wear black," Heidrun says, "Catholics, white." Heidrun's mother's family was Catholic, and her father's was Protestant. As Heidrun points out, "It was not an easy thing to do for a Catholic to marry a Protestant in a Protestant church, but that is what my mother did. It actually made sense to do so, because there were only two or three Catholics in Langenhain."

Confirmation, *Jugendweihe*

Frau Matthies began preparations for Heidrun's confirmation a year in advance, raising turkeys and stockpiling almonds, chocolate, coconut shreds, and all the other required delicacies in short supply. Relatives in the West could be counted on to fill in the gaps. Liquor had to be hoarded as well. There were two dresses to be made—one for the examination in church on the Sunday before Palm Sunday, and the other, the black dress, for the confirmation.

The week before, the baking was begun. Twelve large round cakes, each a different type, were prepared during three days of baking. The Matthies family did not have their own ovens, so Heidrun carried the round tins to and from the bakers on her head. On Wednesday, Heidrun went through the village bearing slices of cake on a tray, calling on friends to announce her confirmation.

On Palm Sunday, the family and godparents went to church. Following the service, they returned for the confirmation meal and remained for coffee and cake and later *Abendbrot* (evening bread). The godparents brought large gifts, linen or silverware for the dowry. Friends of Heidrun's parents and colleagues in the LPG sent numerous smaller presents, for Herr Matthies was well liked in the village for favors he had bestowed on almost everyone. Depending on the size of the gift, people were invited for coffee and supper on Monday or were given parcels of cake. It fell to an embarrassed Heidrun to go from house to house to either thank people for their present or give them cake instead. The following weekend, Heidrun participated in the secular socialist coming-of-age ceremony, *Jugendweihe*.[6] Not all Heidrun's friends celebrated both, especially those whose parents were devout Christians. Some among these friends experienced difficulty later, when making their way in a career. Preparations for *Jugendweihe* were made throughout the year. Obligatory was the trip to the Buchenwald concentration camp in Weimar, where the initiates were lectured on the organized resistance undertaken by Communist inmates.

The *Jugendweihe* ceremony took place in the morning in a large hall, with parents and relatives in attendance. The initiates were called to the stage in groups of five, where they received a copy of the book *Weltall, Erde, Mensch* (*Space, Earth, Man*), presented each year to all candidates. The ceremony ended with the *Jugendweihe* oath, pledged by the initiates, who say, "Yes, that we believe," three times. For *Jugendweihe*, Heidrun cut her hair and got her first permanent.

The last time Heidrun wore her black confirmation dress was for her maternal grandmother's funeral. She remembers standing by the open grave, shovel in hand, worrying whether she had enough dirt to scatter three times like the rest.

School

Heidrun describes herself as an average student during the first five years at the Langenhain polytechnical school, receiving mostly 2s and 3s and only an occasional 1.[7] Her courses included mathematics, biology, physics, chemistry, history, German, and Russian.

In the sixth grade, she began to develop an interest in reading. She remembers going to the local library and taking home books she had heard about. One day, she took out Homer's *Iliad*. The librarian asked, laughing, "What are you going to do with that book?" All Heidrun knew was that she had a yearning to learn. Coming from a family in which there were few books and no one had very much education, she believed she should make use of the opportunity to read everything she could.

For her last four years, Heidrun transferred to the polytechnical school in the larger town of Waltershausen. Heidrun says, "When I was in that new school in that new class with new people, I found out that I was as good as or better than the other students. Once you realize this, you don't want to go back. And that's maybe when I really started studying harder. It was not that I was very ambitious, but once I began to stand in front of the class and hear the teacher saying, 'She did well; she is the best,' well, that is a good feeling."

A course in English was offered in the Waltershausen school in the seventh grade. It was the only foreign language available besides the required Russian. Heidrun enjoyed the class more than any other she had taken. It allowed her to learn about another country through reading its literature. Heidrun says, "I was just curious; I wanted to see and learn and experience more than I could in the village I was living in." The one book she remembers that had a lasting influence on her was *Tom Sawyer*. She recalls her English teacher as one of the best she ever had, but also one of the strictest.

On reaching the seventh grade, Heidrun left the Pioneers and entered the FDJ. Heidrun says, "For most of us, it was more or less a routine, a thing that was expected of us, and you did it unless you had some reason not to, and I guess I didn't have any. You could say no, but then everyone, teachers, et cetera, would try to convince you. You would just stand out. Later, if you wanted to transfer to the twelve-year school (*Erweiterte Oberschule*, or EOS) to obtain an *Abitur* and you were not in the FDJ, then you had little chance of getting in.[8] There were only a handful of students from each class who could go, only the top few percent, and then you had to be not only academically the best but also politically."

To enter the advanced school, Heidrun would have had to apply before the eighth grade. At the time, she had dismissed the idea, not feeling confident that she would be one of the chosen ones. Heidrun says, "I felt it was safer to stay where I was. I didn't have enough information about what EOS meant, how I would go about it. My parents just did not have the time or the energy to help me. They really left all the decisions up to me. It was only later that I began to think about the future."

In seventh grade, Heidrun took the first of four years of the required civics course, primarily an introduction to Marxism-Leninism. Heidrun says, "The trouble was that we only had uninteresting teachers who taught it. They were not very talented, so no one took it seriously. That year's teacher was the least respected of them all. We made fun of him, although we tried, some of us really tried, to understand what it was all about, but the course was not very helpful."

On the completion of her tenth year at Waltershausen, Heidrun began to look about for a way to obtain the *Abitur*. At first she thought of Sundhausen, not too far from Langenhain, where one could learn agronomy and complete the *Abitur* at the same time. More compelling, however, were the recruiters from Zeiss-Jena, who promoted an *Abitur* program in pre-production mechanics as well as employment at Zeiss upon its completion. The program paid all expenses plus provided an allowance of 80 deutsche marks (DM) per month, increasing later to 100 DM, then 120 DM.

In the late sixties, Zeiss was in the midst of an expansion and required trained employees in a number of supporting technologies. Heidrun applied and was invited for interviews, followed by tests lasting an entire day. Heidrun says, "It was my first exposure to a big test. I felt so dumb. I couldn't even spell 'Bertolt Brecht.'" She remembers coming home at the end of that day and telling her mother, "Oh, I don't think they will take me." Only later, when she was accepted, did she realize how much she wanted to go. Returning to Langenhain on her first free weekend, she allowed herself to think, "What would my life have been like if I had stayed here in this narrow world?"

Heidrun was sixteen when she left home and Langenhain for Jena. Neither her mother nor her father attempted to deter her from leaving, allowing her to go her own way as they always had, proud that a daughter of theirs, of peasant stock, was so accomplished in the eyes of others.

Heidrun met Erhard only a few days after she arrived in Jena. Heidrun says, "We were the first group of students to live in the new building. There was a room on the first floor, a room for everyone, a common room. There was a boy who came in with a portable radio, talking to everyone in sight, and that was Erhard. It took a while. We became friends, but I had other friends, and he had other friends. After the first not very favorable impressions of him, I began to be fond of him, but it was not like, 'I want to go with this boy.' It wasn't until the third year that we got to know each other better and to like each other."

Erhard and Heidrun completed their *Abitur* training at Zeiss in the spring of 1971. Heidrun was glad to have it behind her; she had found the program boring. Erhard, on the other hand, had enjoyed it. He had always had a knack for mechanical things. Erhard and Heidrun were the first members of their families to pass the *Abitur*. Even now, Edgar on occasion expresses his pride at his son's accomplishment. "Sometimes it goes to his head," Erhard exclaims. "I couldn't, and still can't, stand that."

Heidrun remembers feeling good about the direction that the GDR was moving in during her first years in Jena. She says, "The society, which was progressive at the time, in the seventies, was developing in a good way. It was good for my life. My personal life seemed to go along with the general improvement in society, so I became more active when I was in Jena at Zeiss than I ever had been at Waltershausen. The FDJ was in charge of organizing competitions among trainees. The biggest responsibility I had was organizing FDJ events at all the different Zeiss departments. It was a big thing and involved students from all over the GDR."

The Sixties

On looking back, Erhard and Heidrun are now of the opinion that most of the experiences of the sixties passed them by. They were so busy with their apprenticeships at Zeiss that they did not notice the scope of it all. For one thing, drugs were not an issue. Liquor was evident everywhere, but it was not an expression of alternative lifestyles. Erhard says that the first time they really learned about it was from the film *Blutige Erdbeeren* (*The Strawberry Statement*), which they had seen in Budapest in 1971, in

English with Hungarian subtitles. They did not really take in the events in Prague either,[9] and it was only through the LPG's partnership with a Czechoslovakian collective that they heard from people who had ripped up their party membership cards. Edgar remembers the large number of GDR military vehicles moving toward Czechoslovakia's border that summer, while he was harvesting hay near the Autobahn for the LPG. Heidrun recalls seeing an interview with Heinrich Böll on West German TV that she tended to discount. At that time, she says, she still had a lot of confidence in her government and believed that it was a counterrevolution that had occurred in Prague.

In the summer of 1971, Erhard and Heidrun went on a trip. Taking the train to Budapest, they hitchhiked through Hungary. At the time, they were not officially engaged; in fact, they had purposely not gotten engaged to avoid, as Erhard put it, "all that stuff." Neither Erhard's nor Heidrun's parents expressed disapproval of their children's trip. Heidrun says, "It is quite remarkable that my parents didn't say anything. I think they trusted us. Erhard's parents, too. They knew me by that time."

Commentary

Erhard and Heidrun were of the last generation to share with their parents and forebears the particularities of coming of age on a working farm. As farm children, Erhard and Heidrun had the opportunity, as had Edgar, Irmgard, and Wally before them, to connect their parents to work and to places that they themselves were familiar with. Not that proximity of workplace meant that farm parents in Göttern, Bucha, or Langenhain had much time for their children; they did not. Edwin and Elly, and in turn Edgar and Wally as well as the Matthieses, were already on their feet in the morning and in the stables when it came time for their children to rise.

It was fortunate for Edgar and Irmgard that Opa Louis was so accessible, as was Oma Thekla when she was alive, to fill in the hours when they were occupied in the stables and fields and with housework. In like manner, Elly bestowed her attention on Erhard, as did Oma Frieda on Heidrun.

Then, too, there was the bustle associated with having so many people about: the hired hands filled all the extra places around the big table in the kitchen where Edgar remembers growing up, just as Erhard recalls being the case when he was a small boy.

There were the chores also—washing the milk cans, bringing in the hay, digging potatoes, carrying and fetching bread to and from the bakers. When the chores were done, there was the experience, shared in common between generations, of finding the village literally at the gates of their respective farmsteads. Not that there was ever much going on in the streets of Göttern, Bucha, or Langenhain in the 1920s and 1930s or in the 1950s and 1960s, but there was probably always something worthwhile to comment on.

The daily routines of farm life continued until the advent of full collectivization rendered the private farm obsolete. By then, Erhard, in his mid-teens, began to steer himself toward a future separate from that of his father and grandfather, just as Heidrun had chosen a different way for herself in contrast to the role of housewife unquestioningly assumed by her mother and grandmother.

The grounds for differentiating between one generation and another had been set in motion by the time Erhard and Heidrun started their schooling in the ten-grade general polytechnical secondary schools in Göttern and Langenhain, respectively—a far cry from the eight-grade one-room schools that Edgar and Wally had frequented in Göttern and Bucha. The schooling that Edgar had experienced was not significantly different from that which Edwin had known when he attended the elementary school in Göttern a generation before him. That both Edwin and Edgar later attended agricultural school only deepened their commitment to farming. The advent of the National Socialists to power did not alter these expectations; it further institutionalized them.

The polytechnical school was the linchpin of the radical program of educational reform undertaken by the GDR. It took the place in the East of the traditional triad of the *Hauptschule*, *Realschule*, and *Gymnasium*, which continued in the West. The polytechnical school made education beyond the elementary level available to children of all socioeconomic backgrounds. It allowed children of farming families, such as Erhard and Heidrun, to go further and faster in their education than would have ever been the case in the customary school system. Attendance at the polytechnical school only confirmed Erhard's decision, made on his own, to not follow in his father's footsteps and become a farmer.

Erhard was a child of the space age. He was excited about the feats of the Russian cosmonauts. He liked technology, and he felt drawn to the claims made by the SED regarding the roles of science and technology in the creation of a socialist society. Once he had decided, at the end of his tenth year, to enroll in the Zeiss *Abitur* program, he entered a new realm of achievement for a member of the Schorcht family.

So was the case with Heidrun; once she had discovered her intellectual potential while at the polytechnical school in Waltershausen, it was only a question of identifying a program through which she could obtain the *Abitur* and then seek admittance to the university. Not only was she motivated to continue her education, the local authorities were more than willing to make it possible for her to do so. Heidrun was encouraged to do what, a generation earlier, Irmgard—victimized by a different ideology—had been prevented from doing.

Erhard and Heidrun represent the breakaway generation. They belong to the last cohort to know what it was like to be brought up on a working farm. They share that "timeless" experience with their parents: grubbing potatoes, hauling bales of hay and straw. And they were the first generation to be propelled away from that tradition to intellectual careers of a different uncertainty.

Erhard and Heidrun worked hard to fulfill the promise of new career possibilities for themselves—in the spirit not of repudiating their parents' lives but of becoming more effective socialists. What they could not foresee as the sixties gave way to the seventies was the possibility that what their generation assumed to be its socialist birthright could itself be the subject of transformation.

16

The University

Seeking Admittance

ON THEIR RETURN from Budapest, Erhard and Heidrun began looking into university programs. At first Erhard thought he might study in Mittweida, in Saxony, but then he changed his mind in favor of the technical school at Ilmenau. He eventually decided in favor of the University of Jena, with the hope that Heidrun would enroll there, too. He applied for the Scientific Equipment, Optics and Electronics program and was accepted.

Heidrun had finished her vocational training with top grades. She says, "I was an outstanding student, but I didn't like it. I could do what they wanted me to do, but I could never excel in that field. I realized that there was something missing in my brain. I could work hard, but I could never make up for that abstract imagination that would tell me that there was a line that I forgot to draw. So I decided that it would not be wise to go on studying in a technical field, and I began thinking about something else; of course, languages came to mind."

Heidrun hoped to find a program in languages, either in translating or teaching. Besides Jena, which had already filled its small quota, only Leipzig or Berlin offered language programs. She did not dare apply to the foreign languages department at the Karl Marx University in Leipzig, feeling that she did not have a chance at a traditional academic program. Rather, she made inquiries at the translator's school, where they asked her, "By the way, where do you come from?" When she answered, "Technical B-Zweig" (the technical option B as opposed to the liberal arts option A), she was told, "You can come and take the entrance exam, but your chances are not very good. There are so many applicants. Think it over."

Heidrun knew that many of those who had applied were undoubtedly twelve-year-school (EOS) students who had three more years of English than she had had. She then sought admission to Humboldt University in Berlin to become a teacher of English and Russian. She was not admitted, being told that Berlin did not accept applicants from the South; students from the South should seek their higher education in that region. "So there I was," Heidrun says, "with no place to go."

Help came from an unexpected source. In the course of her Free German Youth (FDJ) student organization work, Heidrun had become acquainted with officials in the personnel department at Zeiss. When one of them asked Heidrun what she wanted to do, she told him of her difficulties in obtaining admission to the University of Jena. "He said, 'I will talk to somebody.' He must have, for they took me, even though the quota was filled. So actually, I was helped by a connection in the party cadre," Heidrun says with a laugh.

Teacher Training Program

In the fall of 1971, at age nineteen, Heidrun entered the four-year university program of teacher training, leading to a *Diplom* and, ideally, to a position teaching English. In the first years, the emphasis was on linguistics and pedagogy, the theory and history of education, "hours and hours of dull reading," says Heidrun, "and, of course, Marxism-Leninism, with equal amounts of reading." The four-year "M-L" course focused in the first year on philosophy, the second on economics, the third on scientific socialism, and the fourth on an overview and summary. Heidrun says, "It was intellectually engaging, because you had to read the original sources. We read Marx's *Capital*, and it was a challenge to understand it. We really worked hard, with the theory and everything."

There was also a course in literature, Heidrun's introduction to America. The students read Sinclair Lewis, John Dos Passos, Upton Sinclair, John Steinbeck, and Richard Wright. Less attention was paid to English literature. The instructor's specialty was American writing, particularly the "progressive authors," most of whom were published in the German Democratic Republic (GDR) in German. The instructor was like several others who had learned English while prisoners of war in the United States and made a career for themselves in the academic world as teachers or professors of English, with an emphasis on American literature.

In the university, Heidrun's work with the FDJ continued at an even more intense pace. She says, "I got myself a very interesting job as an organizer for summer student exchanges: the Summer Student Brigade. It was my job to find students who were willing to go to Bulgaria or Russia or other places for a few weeks in the summer to work and to receive students in exchange. And another thing I organized was an exhibition of students' work; I enjoyed that."

Heidrun's involvement with the FDJ led her from time to time to think about membership in the Socialist Unity Party of Germany (SED). She says, "I was never pressured to go into the SED. I was approached once by a teacher, when I was still at Zeiss. I was eighteen at the time. I think it was almost routine that they asked. And then I said, 'I don't know. I think I'm still too young. I still have to make up my mind,' and then they left me alone.

"After that I had my *Abitur*, and things were particularly good then, and I began to think, 'Well it's OK, why shouldn't I join the party?' I liked the people who were in the party in the Zeiss school. They didn't look to me as if they were careerists or opportunists. They were really sincere, hardworking, honest people. And I thought I would like to belong to them. Joining up was the last thing I did at that school. I was admitted as a candidate at Zeiss, and then there was a probation period of a year, which I did at the university. I became a full member in the summer or in September of 1972."

East and West

Erhard does not remember thinking that the difference between East and West in the sixties and seventies was so vast. Now when he watches films of that period made in the West, he feels the same way. Both sides, for example, talked about filmmaker Wim Wenders. He admits that he was not in a position to compare the two Germanys, as he knew the West only through TV. By the early seventies, no one his age had been in the West since age five or six.

Heidrun was of the opinion that socially, the GDR was more advanced. She says, "Bonds between people were closer, stronger, in the East. I think that in the sixties, life became more social. Of course, it was official policy to grant social advantages to people, but the social relationships between people were something else—people were not in a competitive situation with their neighbors. You could envy your neighbor's Lada while you had only a Trabi, but it was nothing like today.

"The relative equality of people helped develop this feeling of togetherness. People in the village could know and understand each other, but it was more than that; people felt responsible for each other, they took an interest in each other.

"It also came from the work environment, working in the collective, being on different committees." Heidrun pauses and then says, "I suppose that these social aspects of the workplace could have been some kind of compensation for the fact that other desires were not being met. I mean, certain energies that could not be gotten rid of toward the outside world were invested as positive energy into these groups that they were part of. Still, for whatever reason, people did feel closer back then."

As for how people felt about being part of the GDR, Erhard adds that only older people could assess whether they were GDR citizens. He says, "Heidrun and I were always GDR citizens, but we never had any patriotic or nationalistic feelings like people in the USA do. There were only just a few people who would put out their GDR flag because they were so proud of their country. Personally, I always thought them to be pathological cases." Heidrun ends the conversation by saying that the West had never been important to her, except that it was the country that the packages came from when she was a child.

Marriage

On January 28, 1972, halfway through their first year at the university, Heidrun and Erhard were married. In the fall of 1971, Heidrun had become pregnant. Their friends persuaded them to get married, and they decided on a January wedding. Per tradition, the ceremony and feast should have been held in Langenhain. But, in fact, the Schorcht homestead was larger than the Matthies homestead, and Wally insisted they be held in Göttern.

On the week of the wedding, as tradition dictates, trees from the Schorchts' woods were placed in front of the house and inside the courtyard in front of the door. A large banner was stretched between them reading *"Herzlich wilkommen"* ("Welcome"). Two days before the wedding, a *Polterabend* took place. A cart was filled with crockery and dishes to break. When all had been smashed, everyone came inside for beer, wurst, and schnapps. The last of the company left at 6:00 in the morning.

Heidrun had no intention of wearing the usual kind of wedding dress. She went shopping in Weimar and picked out for herself a long-skirted dress of a creamy champagne color with a sash at the waist. The dress was expensive. Erhard, on the other hand, had to make do with an old suit, as he could not find a new one to fit him properly. The family joke had it that Erhard had to do without.

On the day of the wedding, the bride and the groom led the procession to the church, where the pastor from Magdala officiated. It was the last ceremony ever held there. Years of neglect had led to disrepair, and the building was no longer considered safe.

Erhard had stuffed coins in his pocket to throw at the children as they left the church, "to pay their way from the church to the house." Erhard remembers that his father had told him that he had done that at his wedding, as had his father before him. Both Erhard and Heidrun, who were generally not observant of tradition, felt happy to comply with the old ways on their wedding day. When they arrived at the house, a log barred their way. It had to be cut to allow them to pass to the front door, where they were offered bread and salt before entering the house. The feasting went on into the evening. The following day, Sunday, people who were not immediate family came by in the afternoon to give presents and were welcomed with coffee and cakes and then a meal in the evening.

Wally was complimented on how well she had done for her first wedding. She knew that Norbert's and Roswitha's would soon follow, to say nothing of her own silver wedding anniversary and Edgar's "big" birthdays, fifty-five, sixty, and sixty-five. They were all to be properly celebrated.

On Sunday evening, Heidrun and Erhard returned to their student quarters in Jena. The following week, the Matthieses gave a reception for Heidrun and Erhard in Langenhain. In June, the baby was stillborn. Heidrun says, "Everything went on the way it had before."

Student Life

At first they lived in student dorms, Heidrun in Lobeda in South Jena and Erhard in Zwätzen in North Jena. Then Heidrun moved in with Erhard in his coal-heated room, which seemed no bigger than a closet. In April 1973, they were given a somewhat larger, centrally heated room, 14 square meters, in a dormitory complex for married couples and students from allied socialist countries. The two bunks were made into a sofa by day, while a table large enough to eat from was arranged from two smaller ones. The toilet was down the hall, and the shower was in the basement. Lunch was eaten in the cafeteria, while breakfast and dinner were prepared in the common kitchen. One of the first things they purchased was a small refrigerator, as they found that things left in the common refrigerator were often gone when they went back to get them.

Erhard and Heidrun recall those years with pleasure. They were married, in love, with little to worry about other than finishing their university studies. With their room and board paid for and with a small stipend to boot, they were able to get by. They had no television; few students did, for they were too expensive then. Some evenings they would take in a movie or even a play, or they would join others in someone's room for a party, often talking about books well into the night.

The GDR reading culture was important. Heidrun says, referring not only to their university days, "People read a lot more then. You really had to be informed about new books at that time. In fact, it was hard to have an important conversation without making reference to them. Books were exchanged, passed around. Every part of a book, every allusion was interpreted in terms of what aspect of everyday life was being encoded and commented on. A good book, a critical book, was hard to come by. Of course, not every book could be forbidden, so the way they tried to control circulation was by printing fewer of them. By and large, books were regarded as more important than film, although Ulrich Plenzdorf's *The Legend of Paul and Paula* was an exception."

Their favorite writers—authors of what they regard as classics of the period, which they still keep on their shelves—were Volker Braun (poems), Hermann Kant (*Die Aula*), Brigitte Reimann, Maxie Wander, Helga Königsdorf, and Christa Wolf. They remember reading all of Rolf Schneider's books as soon as they came out and then discussing each one.

They read *Eulenspiegel*, the only satirical magazine published in the GDR, and they subscribed to and read the official party newspaper, *Neues Deutschland*. They continue to read it today, but their style of reading has changed. They used to start from the back with the sports and culture and omit the boring political news in the front, which contained long analyses and official speeches. Now they reverse the procedure and, above all, read the political pages.

As students, whenever they had free time, they were apt to turn on the radio to pick up Radio Luxembourg for Western music, most of it sung in English. In their opinion, there was no real pop music made in the East worth listening to. At official dances, the bands were allowed to play only a small mix of Western music, but if the dance was among friends, it would be 100 percent. "Pearls in Her Hair" particularly stands out in Erhard's mind as a favorite, performed by the Hungarian group Omega (in Hungarian, of course). The first record they bought together was one purchased in Budapest; regulations were looser in Hungary.

Heidrun remembers the joy she felt on finding a Simon and Garfunkel record, quite by chance, in a record store in Jena. Someone had hidden it among popular German numbers. She purchased it immediately. A customer was allowed to buy one Western remake record at a time. Heidrun and Erhard played it repeatedly on weekends in Göttern on the Schorchts' record player.

Erhard's perception of socialism in the late sixties and early seventies is relatively positive. When he looks back now on those earlier years, he says of them, "There was a period after the Wall went up when things slowly got better and better. The West was not that much further advanced. You can see that in the films dating from the beginning of the seventies. Sure, there were exceptions, but in the sixties, things were going pretty well here. With a particular education, you had the feeling that you could do something. There was talk of the 'all-around well-educated socialist person.'

"In Göttern, nothing had been destroyed during the war, and things went on as they had before. In the city, we noticed that here and there a street was being built and buildings were going up. It went slowly, but there was a sense of headway, of moving forward. You could believe [Walter] Ulbricht when he said, 'Overtake capitalism without adapting to it.'

"Then things began to change in the seventies. You can't always tell when it started to worsen, but I think it began when Ulbricht was replaced by [Erich] Honecker. Economic problems that didn't seem to be there before began to appear. By then the running joke was, 'We are leaping over capitalism; right now we are kneeling to gear up for the jump.'"

International Student Brigade

For three consecutive summers beginning in 1974, Erhard and Heidrun participated in the Summer Student Brigade that Heidrun had been organizing for the FDJ. It took them to villages in the Soviet Caucasus, all expenses paid, to build houses. They lived together in a camp with Russians their age. As Heidrun observes, "Coming from a peasant family, I know what work is like; I wanted to find out what work was like in Russia." At first, their years of Russian in school and university did not stand them in good stead, but within weeks they were conversing with relative ease. Heidrun describes it as a wonderful experience. She found the people to be exceptionally friendly and touchingly generous with what little they had to share. The Russians their age besieged them with questions about life in Germany, about clothes, about music, about "beat and rock and roll." There was very little discussion of politics and of peace. "It was somehow taken for granted," Heidrun says, "that we were on the same side, so to speak."

Erhard and Heidrun found that they were envied as Germans. The Russians they met liked Germany and wanted to go there. Heidrun says, "They would tell us, 'You Germans have everything, so much culture.' German culture was always something mystical in their minds." It came as a surprise to Erhard and Heidrun that the Russians they spoke with did not seem to harbor any resentment toward Germans. One experience in this regard stands out clearly among the many memories Heidrun has of the Soviet Union. The city of Mineralnye Vody, situated in the flatland in front of the Caucasus Mountains, had been the scene of bitter fighting between Germans and Soviets. While there, Erhard and Heidrun met an old man who said, "We liked the Germans when they were here. We want them to come back." Heidrun remembers having to catch her breath; she could not believe what she had heard. "We met quite by chance one Saturday afternoon," she recalls. "He sat there on a park bench next to us. He had heard we were Germans, and he wanted to compliment us. He said that they wanted the Germans to come back, saying it was much better than what they have now!" Heidrun found that she was flattered and disconcerted by what he said. Erhard was not certain about the man's motives, but he could not help but agree with the intent of his remarks.

Erhard asserts that he was "shocked" at how primitive the people's standard of living was when he visited the Soviet Union for the first time in 1974. "It was as if they were still living in the nineteenth century in the villages," he says. He came there with the image of spacecraft and cosmonauts in his mind's eye. What he saw was starkly different. He likened it to abandoning a car on the side of the road: "Once the windshield is gone, the rest deteriorates very quickly; that's what it is like in Russia. Things are left to fall apart."

Heidrun's Thesis

In her last year of university, Heidrun wrote her thesis for her *Diplom*. She wanted to write about American authors but felt unprepared. Instead, she chose the subject of universities and colleges in the United States, which grew out of her sense of there being important differences between German and American higher education. Her focus was on how American colleges and universities are organized and how they function. What interested her most was the freedom American students seem to have to fashion their programs of study through electives and combinations of majors and minors, options not available to the same extent for East German students. Heidrun had to scour every possible source in Jena and went to Berlin as well to locate information. Her final product was well regarded by her readers.[1]

By the time she was granted her *Diplom* in 1975, Heidrun was on her way to becoming an Americanist in a country where the United States of America was regarded as the principal source of militaristic expansionism and capitalistic imperialism.

Erhard also received his *Diplom* in 1975. It was assumed that as Erhard had trained at Zeiss, he would be taken on there; such was his intention. Just before Erhard signed the contract, however, the chairman of his department approached him, asking whether he wanted to be a research assistant in the Department of Technology. After some vacillation, he chose the university.

In Heidrun's case, what her next step might be was less clear. There were positions available for a person with a *Diplom* to teach English, and although Heidrun was qualified, the thought of graduate studies appealed to her more than teaching did. Graduate work was made available, however, by invitation only. Fortunately for Heidrun, faculty members of the department came forward, inviting her to continue her studies. She was assigned a professor to work with on her dissertation and was relieved of all course work except for the inevitable classes in Russian and Marxism-Leninism. She was expected to complete her work in three years. Heidrun found graduate work to be a lonely undertaking, in which all the responsibility lay with the student.

Heidrun's Doctoral Dissertation

It was fortuitous for Heidrun that an American professor, Dr. Edgar Schick—an administrator from St. John Fisher College in Rochester, New York—came to the University of Jena in the fall of 1975. Heidrun learned of his coming from Professor Peter Schaeffer, an Americanist at the Institute of Comparative Higher Education in Berlin and an acquaintance of Professor Schick. Heidrun had met with Professor Schaeffer in the course of her thesis research, and he had subsequently written Professor Schick about her thesis on American colleges and universities and her desire to write a doctoral dissertation. Heidrun already had a topic in mind—the American student movement, suggested to her by her thesis advisor at Jena, Professor Erich Leitel. The suggestion was more than purely academic, for it was a widely held opinion that the student movement constituted the beginning of an American revolution—the heralded downfall of its capitalist imperialism. Heidrun wondered whether Professor Leitel appreciated how nearly impossible it would be to research the topic, given such limited resources.

On Professor Schick's arrival in Jena, a meeting was arranged with Heidrun to talk about her dissertation. Dr. Schick told her that now that she had written her thesis on the organization of American higher education, she was in a good position to look into the American student movement. He asked her to send him a draft of her first chapter for his review. By the time she left the meeting, Heidrun knew the topic was right for her. Once it was in her head, she could not put the thought down, even though she had no idea where it would take her.

Before she could begin to apply herself full-time to the research, she was required to finish her studies in Russian. Already fluent orally, she now had to show her aptitude in reading and writing. Almost a year passed before she was able to take the final exam. She was also required to enlist in a two-year course in Marxism-Leninism, attending classes three times a week. The focus again was on philosophy, economy, and what was called scientific socialism, similar to what Heidrun had taken as an undergraduate. According to Heidrun, what the M-L department wanted students to get from the course was the inevitability of socialism as the necessary step forward in the development of mankind. "I tried very hard to understand, to make sense of the theory and the reality and the theories within themselves," Heidrun says, "and because I tried so hard, I was never among the best. I always achieved 'good' but not 'very good.' I just couldn't work it out.

. . . It just didn't come together for me. If we had been allowed to pursue the material that interested us, it might have been different, but the topics were set by a professor or even a higher board of people who said this or that is what is to be discussed." Marx was used for economic theory, Lenin for "recognition theory" and confrontation with power. Only later, when she drew on Marx for her doctoral dissertation, did Heidrun come across references to the earlier humanistic manuscripts.[2] They had not been brought up in the course.

The problems Heidrun had in researching material for her undergraduate thesis were compounded in the case of her doctoral dissertation. "We were very innocent," Heidrun says. "We had no way of comparing what we did, in that kind of research, to what people in America or other Western countries did." There were Russian books in Jena on the subject of the American student movement, but apart from some useful references, Heidrun had doubts as to their reliability, largely because in the communist world of the East, the American student movement was an approved topic, a sign of "capitalism in crisis."

Heidrun spent innumerable hours commuting to Berlin, searching for material in the city and university libraries and consulting with Professor Schaeffer. The most useful material, however, was found in books and articles sent to her from Rochester by Professor Schick. The more she read, the more she realized how much she did not know. Just the vastness of America, with manifestations of the movement taking place in colleges and universities so far removed from one another, was, in and of itself, intimidating.

As 1978 approached, the year her dissertation was to be finished, it was, in fact, far from completion. She had written the introduction and the first two chapters and sent them off to both Professor Schick in Rochester and Professor Schaeffer in Berlin for criticism and advice. She realized from their comments that what she could accomplish with the materials available to her would be acceptable in the GDR but nowhere else.

The Birth of Maria

By July 1977, Heidrun knew that she was pregnant. The year ahead promised to be one of change for her and Erhard. One thing was certain: with a child on the way, they would have to move out of the room they had been living in since 1972. At first, they looked for an apartment in Jena, but without success. Then they thought of buying a house, with an apartment to rent to cover the mortgage costs, but no houses were available.

Just then Erhard's parents said, "Why not move out to Göttern? We have all this room." Erhard and Heidrun agreed, as long as they would be allowed to pay the rent. "No, of course not," was the answer. "Electricity and heat, yes, but no rent." They made the move in the late fall of 1977. Erhard and Heidrun took over the bedroom, the living room, and the small kitchen on the second floor. At first they all ate together in the big kitchen downstairs, where Erhard's mother did the cooking. Before long, Erhard and Heidrun found that their tastes in food, having changed since living in Jena, no longer agreed with those of the older people. Furthermore, Erhard's parents did not wish to eat their evening meal as late as the younger couple was accustomed to. When Heidrun decided to do her own cooking upstairs, Erhard's parents seemed relieved.

Maria was born in the Jena University Hospital on January 4, 1978, marking the beginning of Heidrun's maternity-leave year, her *Mutterjahr*. Now she had a legitimate reason not to finish her dissertation on schedule. Certainly no one expected her to take time away from caring for her baby to work on it. At first it seemed strange not having to rush off to school or the university, but soon she noticed how readily her day was taken up with Maria's care and all the things to do around the house. Erhard's mother was out of the house a good part of the day, working in the agricultural production cooperative (LPG) cow stables in Magdala. Erhard left first thing in the morning for his job as an assistant in the technology project department at Zeiss, while his father spent the day in the mayor's office.

At the end of the year, Heidrun placed Maria in the day-care center in Magdala. She brought the baby every morning on the 7:30 school bus, left her at the center, and caught the bus to Jena, arriving at the university by 9:00. In the afternoon, her father-in-law would pick Maria up in the Trabi, so she would already be at home when Heidrun returned at 5:00.

The English/American Studies Department

At the start of 1979 and at the conclusion of her *Mutterjahr*, Heidrun signed a four-year contract with the University of Jena to assistant teach in a new area-studies program, Great Britain and the United States, under the chairmanship of Professor Leitel, who had introduced her to area studies (*Landeskunde*) as an undergraduate. More acceptable to the authorities, however, was the title, Research on Imperialism (*Imperialismusforschung*). Heidrun was given a desk in an office in "the Tower" (a tall circular building, meant to resemble a telescope, housing mostly the university administration) with a woman who had been teaching area studies from its inception just two years before.

Heidrun's salary was 760 deutsche marks (DM). Between them, she and Erhard were bringing home 1,500 DM per month. Their largest expenses were food, clothing, and transportation. It was, they found, difficult to save. "We always enjoyed life," as Heidrun says. They liked having people in for coffee and cake or beer, or less frequently a meal. When they did manage to put a little money aside, they bought some new furniture and had cupboards and shelves made in the living room, according to Erhard's specifications, for their growing collection of books, records, and tapes.

The Invitation to America

It was April 1979 when an invitation, sponsored by the International Visitor Program of the U.S. Information Agency, came for Heidrun to visit the United States. Professor Leitel told her about it. At first it seemed like a dream. As far as she knew, such a thing had never happened to a student before. In fact, it was unprecedented. Heidrun Schorcht, age twenty-six, was the first in the social sciences to be so honored. Only Professor Leitel had been allowed to visit the United States before. What made it all the more surprising was that, for security reasons, no exchange students from the United States had ever been allowed to visit Jena, the site of the Zeiss Optical Works. Heidrun tried to keep her expectations under control. She knew she was probably in for a long bureaucratic wait.

From time to time, Heidrun would hear of someone she knew being questioned by the Stasi in regard to the proposed visit to the United States. She believed she had nothing to cover up, and apart from her father-in-law's sister, Irmgard, living in Pirmasens, she had no connections in the West.

There was no way of knowing what the Stasi would do. Even in cases where she knew people to have been devoted members of the party, they were nevertheless not permitted to travel. When she went home on one occasion to visit her family in Langenhain, her father said, "Daughter, what did you do? Have you done something wrong?" He was referring to the fact that the Stasi had been to his place of work in the LPG, inquiring about him, his ways, his opinions. Normally Heidrun's father would not have known about the inquiries, but subterfuge was difficult to maintain in the LPG. The cadre leader had told him that the Stasi had been asking about him. She had to reassure him that it was because she wanted to visit the United States.

Approximately four months after she had first heard from Professor Leitel about the invitation, she received notification from the university's Department of International Relations that she would be allowed to travel to America. Heidrun had been cleared at the highest security levels. She was regarded as a GDR citizen of good standing in the eyes of the party. In addition, leaving a husband and a young child behind ensured that she would return to the GDR.

Now all that remained was to obtain an entry visa from the U.S. consulate in Berlin. As GDR citizens were not allowed to enter the consulate, it was necessary to proceed through the Ministry of Higher Education. Heidrun submitted her passport to the ministry's Office of International Relations, which in turn forwarded it to the U.S. consulate. Weeks of further delay ensued. When her passport was duly returned to her with the visa affixed, close to half a year had passed since she had received the invitation.

Heidrun's departure was set for October 6. She was to fly from Berlin to Prague and from there to New York City. Her itinerary in the United States was worked out through correspondence with Professor Schick in Rochester. For some time, it had been clear that it was he who had initiated the invitation to Heidrun through contact with the University of Jena. Her itinerary, which was to start with a visit to Rochester, was organized to help her collect material for her dissertation.

The days before she was scheduled to leave seemed interminable. She could not help but fear that something would occur at the last moment to prevent her from going. On the eve of her departure, Erhard drove Heidrun and Maria to Langenhain. Maria, then twenty months old, was to be left with her grandmother for the nine weeks that Heidrun would be away. Heidrun's sister, living with her mother, had a daughter two years old, so Maria would have a companion. Heidrun felt a tug at her heart when she left Maria in her mother's arms. She had no concerns about her well-being, only that her daughter might be estranged from her when she returned.

The following morning, Erhard drove Heidrun to Jena, where she took a train for Berlin. From Schönefeld Airport, she flew to Prague and there boarded the Russian-built plane that was to take her on the thirteen-hour flight to New York. Only when the plane had cleared the Prague airfield did Heidrun feel, with uncertain anticipation, that she was finally on her way to the United States, the bastion of imperialism.

Commentary

In 1968, when Erhard and Heidrun were seventeen and sixteen years of age, respectively, they left home to obtain the *Abitur* and subsequent university education in Jena, all expenses subsidized by the state. Quotas for children of peasants and workers were established in the 1950s for admittance to the university. In this regard, Mary Fulbrook points out, "By the mid-1950s, 53% of the university students in the GDR were from the working class (which constituted 69% of the population), in contrast to only 4% in West Germany in 1950 (when 57% were deemed to be working class), rising to a mere 7.5% in West Germany by 1970."[3] Later, preferential acceptance of working-class offspring gave way to encouraging talent regardless of social background. There was a marked emphasis from the middle of the 1960s to the middle of the 1970s on higher education. These years witnessed an urgency in furthering technological development in the more highly industrialized nations. In East Germany, the loss of professional persons to the West prior to the building of the wall in 1961 lent a particular importance to the effort to refurbish the pool of educated talent.

On Erhard and Heidrun's return from a summer trip to Hungary in 1971, they took up residence in a university student dormitory after their marriage in 1972, following the news of Heidrun's pregnancy. At this point in their lives, the distance they had traveled from their beginnings in farming communities became most clearly pronounced—a distance that was testimony to the opportunities offered to them as members of the working class by a socialist GDR.

Edgar, Wally, and Irmgard came of age during the Third Reich, which assumed a different outcome for members of their social class. Elevated to a status of mythic proportions, that of idealized German peasantry, they were irrevocably anchored to the soil. Social mobility was proscribed, entrusted as they were with tilling the "soil" and rejuvenating the "blood" of the *Volk*. At the age when his son, Erhard, received his *Abitur*, Edgar was already fighting on the Eastern Front to help conquer the "living space" Adolf Hitler deemed necessary for colonization. Edgar in 1942 and Erhard in 1972 had arrived at markedly different places in their lives. The thirty years separating the two generations signaled a watershed in German history. Edgar's generation was witness to the defeat of the Third Reich and its antiurban ideology, participating in the birth of a communist East Germany dedicated to the industrialization of the countryside.

Edgar brought his working life to a close, turning from his birthright in agriculture to politics. Erhard and Heidrun began theirs by moving from the farm to the city to embark on careers in teaching and research. Then in 1977, something anomalous occurred, seemingly inconsistent with the projected industrialization and urbanization of socialist life. Erhard and Heidrun joined Edgar and Wally to live together in the family house in the village. Some of the historical and social distance separating the older from the younger generation was thus shortened.

17

Heidrun in America

The Dream Comes True

EIDRUN RECALLS landing in New York: "It was overwhelming. When I arrived in Kennedy Airport, I was so tired, I was wide awake. I had the feeling that everything was unreal." The person who was to meet her from the Foreign Visitors Office was not there, so she took a cab with two others also needing to go to La Guardia. Heidrun sat in the front seat next to the driver. It was a beautiful, clear, starlit sky with a full moon. "It was," she says, "a surrealistic scene—easy music on the car radio, with me in the front seat floating along to who knows where. The whole thing was like a dream."

Professor Edgar Schick was at the Rochester airport to meet Heidrun's USAir flight. At the Schicks' home, Heidrun met his wife and family and immediately felt at home. She spent the first few days with the family, catching up on her sleep and further planning her itinerary. On her first morning in America, she was introduced to pancakes and maple syrup.

Her initial archival work was done in the library of St. John Fisher College in Rochester. Then she moved onto several colleges in upstate New York, where Professor Schick knew there to be files. In one, she met her first former student activist, a man who had been involved in the movement. For the first time, she felt that her research was beginning to take on a reality that it never had before.

She found that she could not get over the scenery of upstate New York—the hills, the woods, the lakes. Everywhere she looked, it seemed as if the trees, in the full peak of their fall splendor, were on fire.

She stopped over in Chicago to see the famous buildings she had heard about and then made a visit to the library at Ann Arbor before heading westward toward Madison, Wisconsin, the goal of her trip. She spent a week in the archives of the Wisconsin Historical Society. Boxes and boxes of SDS (Students for a Democratic Society) files were piled everywhere. It would have taken months to go through them all, whereas she had only a few days and limited funds for photocopying. Frustrated, she confined herself to copying the most important documents and taking notes on the rest. By the end of the week, she knew she had the data with which to write an acceptable dissertation. She threw away the first three chapters she had written in Germany.

Everywhere Heidrun went, she found people interested in helping her with research on the American student movement, as if the knowledge about that assault on the American establishment was to be made available rather than sequestered away.

From Madison she flew to Washington, DC, to make use of the Library of Congress. There she marveled at its size and the extent of its resources. "It was rewarding to be there," she said, "just to be there." The Foreign Visitors Bureau helped Heidrun see the city and its surroundings. "I took advantage," she says, "of the inexpensive tour to Mount Vernon. I was able to visit Congress, to sit there in the gallery of the Senate and the House of Representatives, and to visit the White House, the home of the president of the United States, which belongs to the people."

Before Heidrun left Washington, she made a special trip to the National Archives to view the American Declaration of Independence and the Constitution, to reflect again on what they had to say concerning the rights of the individual in a democratic society. Later, she read Henry David Thoreau's treatise on civil disobedience.[1] Nothing else made more of an impression on Heidrun than Thoreau's claim to the individual's right to dissent.

The longer she stayed, the more confident she felt about her English and her ability to meet with strangers and talk. She says, "I met [Eugene] Genovese once. That was organized by Edgar. I had a nice talk with him and his wife. It was maybe two hours. At that time, of course, being so ignorant, I could not really appreciate who I had there—I had not read any of his works before. That was the case several times—that I was in places or met someone and had no idea how important they were. That was because of the nature of our studies at home; they were so limited."[2]

Even though Heidrun found people to be kind and helpful, most had little interest in East Germany and very little knowledge of it. In time, Heidrun began to feel that people did not hear her when she said that she was from East Germany; rather, they assumed she was from the eastern part of West Germany. "If I didn't point it out," she says, "they would not have noticed. I'd say I am from the East, and they would say, 'Which city?' and I would say, 'Jena.' No one had heard of it; Leipzig, the same thing. Some had heard of Dresden. Then people would ask if I wanted to stay in the States; if I said no, they would want to know why. I'd say, 'I have my family there.'"

She recalls an incident at Thanksgiving at the Schicks': "They had invited friends, people Edgar knew from church who had just moved to Rochester and didn't have any family there or friends. We happened to talk about wine over dinner. The man asked me if I liked Rhine wine. I said that I had not had any Rhine wine but enjoyed Hungarian wine. Then the man said, 'How can it be that you are German and you don't know Rhine wine?' I told him that I was from East Germany. For a while he was quiet; he didn't know what to say. Then he asked, 'Well, how come you are here then?' I had to explain that Edgar had procured an invitation for me and that for some reason they had allowed me to leave. But it was too much for him."

Heidrun wished she would have had more opportunities to talk to people. It was partly due to the fact that she had spent so much time in libraries, and then, too, she says, "I am not the kind of person who is outgoing or who makes contacts easily. I felt lonesome at times, because I couldn't talk to people. I would have liked to [have spoken with] somebody, but I just couldn't go to a stranger and say, 'Hey, I want to talk to you.'"

Heidrun had never seen anything like an American suburb before she became familiar with the one in Rochester where the Schicks lived. "All those single American family homes," she exclaims, "each one with a lawn in front and behind, and a garage and no fences anywhere. You could walk through the streets and see these neat houses everywhere, all very well kept, and nothing else—no center, no village, except for the supermarket! The first supermarket I saw was a Wegmans—that was something! The wealth, the material wealth, impressed me. But as for all the negative things about American that I had been made aware of at home, like the slums, well, I didn't see any. I spoke to a black man in Rochester, and he said that there were slums, black areas, but I didn't get to go there. When I was in New York, no one took me to the Bronx or to Harlem, which we had all heard so much about. I was always looking for some real slums to take pictures of. I really felt I should, but I didn't see any. The closest I came was in Philadelphia. Edgar said, 'There is this part of town, but we won't go there. One of us would get killed and the other raped, and you can guess who that would be.' When I heard remarks like that, I knew something was there, but I never saw it myself."

Toward the end of her stay, Heidrun began to realize how much she enjoyed the American middle-class way of life that she was being exposed to, especially as she experienced it with the Schicks. "It was the little things," she says. "When Edgar came home, he and Peggy had a drink before dinner and sat there and talked a little. I liked their parties, receptions, and the way Sundays were organized: church with coffee after the service, and the comfortable coming and going of neighbors and friends."

When the time came to leave, Heidrun found it harder than she thought it was going to be. "I had the feeling," she says, "when I said good-bye that I was probably seeing these people for the last time in my life. I hate final things. It was depressing."

The trip back, as Heidrun describes it, was difficult. "When I left New York, it was raining, and I sat in this small glass cubicle that the Czech airlines had for their counter. When we landed in Schönefeld in Berlin via Prague, it was dark, very dark, compared to New York with all its lights. We had to go to the old train station there that is like a barn—dirty, cold, uncomfortable. The contrast to where I had just been was very strong. In the compartment on the train was another person who, it turned out, was very interested in the United States. He was so surprised to find someone who had just come from the States who spoke English. He was German, an engineer from Jena or Apolda. So I gave away my *New Yorker* to him. I was so happy to find someone who was interested in the United States. I was still clinging to my stay there."

When Erhard met Heidrun at the station in Jena, they drove directly to Langenhain. They arrived shortly before dinner. When Maria heard the car pull in, she came running out of the house as fast as her short legs would carry her. "I said, 'Hello, Maria, I am back.' Maria took my hand and led me into the living room and said, 'Mummy eat, mummy drink.' It was as if I had never been away."

For a while after she came back, Heidrun attempted to keep alive the flavor of American life she had grown to like. She brought back several recipes from the Schicks' kitchen and began preparing American meals— hamburgers, pizza, apple pie. Most precious, though, was the maple syrup, which she served to her family with pancakes on Sunday morning. To make the American atmosphere as authentic as possible, she bought a new refrigerator and introduced the family to ice cubes. No matter how hard she tried to follow the recipes, however, whatever she prepared did not taste the same as it had in Rochester. Little by little, Heidrun allowed America to slip away.

Never had she been busier, with classes to teach, her dissertation to write, and a baby daughter to care for. She found the following two years, 1980 and 1981, to be her most rewarding: "It was the most intense time in my life."

Erhard behind the Camera

The job that Erhard had taken in 1978 with Zeiss turned out to be more interesting than he had anticipated. The largest project he worked on, always in association with others in groups of three or four, never came to fruition, but he found it exciting nevertheless. "It had to do," as Erhard describes it, "with the construction of a factory to produce microelectronic chips. We designed everything from the ground up. It was to be a huge factory, built in Burga near Jena. The building was already designed; we had just to equip it, on paper—everything—water, gas, electricity, a dust-free environment. We even had a few physicists who worked for us. They would always come up with new discoveries. The things we came up with couldn't, however, be too modern, because we had to order the machines, and they took a couple of years to build; meantime, new stuff was being discovered. But we were working in virgin territory. We were doing what the Japanese were doing. How it would have all turned out, I am not sure, but for financial reasons, it never got finished." With the termination of that project, Erhard's position at Zeiss came to an end in May 1980. Then, quite out of the blue, came an offer from the university's media department: Erhard would make films for training and documentation purposes in the fields of medicine and sports. It involved high-speed filming to show technical and surgical procedures in slow motion. The job connected with his boyhood interest in cameras and taking pictures.

When he was twelve years old, Erhard had saved up 16 deutsche marks (DM)—8 DM in big coins and the rest in small change—and bought himself a 6-inch-by-6-inch medium-format camera. "Something like the Trabi," Erhard says of it. Erhard later acquired his father's camera, which didn't work anymore, taking it apart to see how it functioned. He also inherited an SLR camera with an internal light meter from his grandfather.

In 1968, he became a member of the Jena amateur film club and began making 16mm films with other members. The club received apparatus and film stock from Zeiss for free. The experience with the club as well as an apprenticeship at the Zeiss Film Studio stood him in good stead for the next job that opened up.

In 1985, Erhard joined the university's technical department for automation and rationalization, where he worked with a multispectral projector, a machine used to evaluate pictures taken from outer space. Six cameras with different filters took pictures in different wavelengths: two in the infrared spectrum and the others in visible spectrums with different lenses. With this projector, Erhard could evaluate the pictures of Earth by mixing together various photos using the six channels of the projector. In this way, he was able to make "fake colors" that could be used to identify more clearly certain features on Earth: water, oil, pollution, and so on. It was a three-tier system that incorporated photography from space, aerial photography, and pictures from the ground. As Erhard describes it, "Our job was to become familiar with this machine and its potential in order to be able to find new applications for it, such as in architecture, stress tests, and so forth."

Nevertheless, the project was not considered a success. It was undertaken to fill a gap, to produce high-tech units not available to the German Democratic Republic (GDR) because of the embargo. When they did begin to hit pay dirt, it was too late. Most of the products they hoped to produce became available on the market soon after unification. Erhard believed that much of what he had done had been in vain.

Writing the Dissertation

When Heidrun began to write her thesis, it came slowly. She found that being in the United States had so widened her perspective that her earlier generalizations were no longer valid. One thing had become clear: the need to ground her dissertation in American history. The more she pushed her analysis, linking the student movement with instances of other protests in American history, the clearer it became that the events of the late sixties were part of a long American tradition of dissent and change. As this new realization took hold in her mind, Heidrun felt ebullient at understanding an aspect of American history that neither she nor members of the faculty at Jena had grasped before.

All through the writing process, Heidrun had the support of Professor Schick. She would send him chapters, and he would comment and send them back. Heidrun says, "He would say things like, 'That is fuzzy' or 'I don't see your logic here. Back this up if you can.' But he never tried to alter my ideological slants, so to speak." The fact that he had been in the administration of St. John during the protests, and thus had the perspective of the other side as well, added credibility to Heidrun's work.

One of the major changes in her thinking that emerged as she wrote was that the American student movement of the sixties had far less to do with economics, as she had originally thought, than with attitudes and the ways in which decisions should be made. What was radical in Heidrun's work was not that she was locating an American revolution in the student movement but that she was diverging as widely as she was in her perspective from the accepted canons of thought in the party. She knew that she was not going to make a complete break with those modes of analysis, nor did she want to, but she did wonder how her colleagues who taught Marxism-Leninism would react to what she was doing.

When the dissertation was done, Heidrun had someone else type it: She did not trust her typing, nor did she have a word processor. For the twelve copies she was required to produce, she used offset printing. "There was a joke," she says, "that even the doorman gets a copy. It took a lot of time, energy, and money, but I wanted to do it." She adds, "The family was very supportive. My father-in-law drove me all the way to Tannroda (some 20 kilometers to the west of Göttern), because there was a printer there who did offset printing."

Heidrun describes her dissertation in its final form as follows: "It starts out with the typical thing about an anti-imperialistic outbreak, but the thesis itself is that the student movement was a reform movement in the American tradition of a struggle for civil liberties and freedoms that can be traced back to colonial times and the Declaration of Independence. It did have an extreme left-wing dimension, but that was not the dominant focus. It was much less about the economy than it was about the basic needs of people. It was part of a tradition of protest movements that aimed to improve the system but not to overthrow it."[3]

There was a large turnout for Heidrun's dissertation examination itself: the three readers—Professor Schaeffer from Berlin; Professor Karl-Heinz Schönfedler; and Professor Leitel, her earlier mentor—as well as the members of her own department, Landeskunde USA/Grossbritannien; representatives of the staff of Marxism-Leninism; and Schick, who flew over for the occasion. "That was something," Heidrun says. "I was the first one to have a professor come from so far away." Professor Schick was regarded as an unofficial fourth reader whose opinion held weight.

The examination went more smoothly than Heidrun had thought it would, with less resistance to her thesis of reform rather than revolution than she had expected. Even the delegation from Marxism-Leninism refrained from making an outright attack on her, confining themselves to questions about the American working class and the lack of reference to its role in her dissertation. Heidrun says, "I pointed out that the working class didn't play a part in the movement I studied, except for a negative one. There were elements of the working class that were opposed to the student movement." The dissertation was approved without dissent. Heidrun felt relieved and a little let down. She had invested so much of herself in the research and the writing, and now it was over.

Commentary

Following her trip to the United States and the completion of her dissertation on the American student movement, Heidrun began to feel like a true Americanist. She had paid her dues. She had stared the monster in the eye and seen reflected there her hopes for a better German socialist society. The developing relationship she had with the United States was complex.

It had begun with reading Mark Twain's *The Adventures of Tom Sawyer* and being charmed by the evocation of childhood freedom in a new world.

Much of what she read at the university by Dos Passos, Wright, Hemingway, and Bellow was about lives lived at the edges of convention. America began to become for her a place where people appeared to have a degree of freedom of choice unequaled in the world she knew. This theme was reiterated in her undergraduate thesis on American colleges and universities, in which she noted how American students had the possibility of shaping their own curricula through course selection to a degree not known in the GDR. Such a "progressive" perception of American life contradicted the general understanding of the United States as the imperialist bully in the world arena, the enemy of people's movements everywhere, whose military might was a threat to human life. This ideological "reality" of America predominated in Heidrun's thinking about the United States, one that she had first become familiar with in the Ernst Thälmann Pioneers in Langenhain, later again in the Free German Youth (FDJ) at the polytechnical school in Waltershausen, and then as an article of faith in the university chapter of the Socialist Unity Party of Germany (SED). It was a view she found elaborated on almost daily in the party newspaper, the *Neues Deutschland*, one she shared with friends and colleagues whose opinion she respected and that was reinforced by her contact with them. It persisted even as the trip to America led to a crisis of confidence in the party's cause.

The other view of the United States, which had to do with freedom and with openness rather than with oppression and domination, was one that was revealed to Heidrun in her reading in Jena, first in her English literature class and then in preparation for her thesis. It had come to her inadvertently, almost as an unwelcome guest in her mind. There were few persons other than Erhard with whom she could share it.

The choice of her dissertation topic, the student movement in the United States, could serve only to aggravate her dilemma, even though it was not necessarily experienced as such by Heidrun. The subject suggested to her by her adviser, Professor Leitel, was regarded as auspicious for a candidate for a doctorate in imperialism research. Heidrun willingly accepted the theme, but perhaps for different reasons than those her mentors had in mind. It may have been her wish to identify with a cause she and Erhard believed had passed them by. Only later, they said, did they appreciate the significance of the scope of the movement.

America held a number of surprises for Heidrun: the beguiling quality of American middle-class life, the beauty of the autumn landscape of upstate New York, the lack of evidence of grievous injustice at every hand, the strength of democratic institutions in American community life, and, not least, the American Constitution. Indeed, it is fair to say that Heidrun found in the ways of American democracy an answer to the perceived "missed opportunities" necessary to transform East Germany into a democratically organized socialist society. Yet as persuasive as was Heidrun's witness to this meliorative aspect of American life, she remained loyal to the idea of America's expansionistic designs, against which the Soviet Union's ballistic armory was justified. The Cold War was the truth of their lives, and she knew on whose side peace was.

But if Heidrun's perception of the United States was ambivalent, so was her view of the Soviet Union. Russia was the defender of the communist faith and the first friend of the GDR, but unlike that of most best friends, the German-Soviet friendship was an institutionalized obligation of the GDR citizen, which, as such, called the relationship into question. Heidrun and Erhard traveled to the Soviet Union in three successive summers to help the Soviets build housing. What they received in return were gratifying personal friendships, while at the same time they witnessed restrictive absolutisms and a demoralizing standard of living.

The ice was to be finally cracked, if not broken, for Heidrun, not by anything the U.S. government did or said, nor by any gesture or action made by the Soviet Politburo to break the intransigence, but by the dissident voice of the GDR peace movement that the American democratic tradition allowed her to hear. It was the voice of the Berrigan brothers,[4] saying that peace was indivisible, or that of Henry David Thoreau, or Elizabeth Cady Stanton, or Frederick Douglass speaking in defense of the individual's conscience, voices that for Heidrun could remove the pretext of ideological "truth."

18

Heidrun's Metamorphosis from a Cold War Kid

Teaching the American Constitution

W HEN HEIDRUN'S DISSERTATION was accepted in 1982, one year of teaching remained in her four-year contract. By then Maria was four, with long blonde hair like her mother's. Heidrun wished she had more time to spend with her daughter, but the stress of teaching, doing research, and writing made for full days during the week. She managed to spin off a couple of papers from her dissertation for the one journal in her field, *Zeitschrift für Amerikanistik und Anglistik (Journal of American and English Studies)*. The waiting period was two years from the time of acceptance until publication. An easier way to appear in print was to present a paper at one of the frequent conferences dedicated to research on imperialism, the proceedings of which were often published. When she attended, Heidrun would join the work group on the United States so that she could meet colleagues and exchange information and ideas. In plenary sessions, prominent professors of imperialism would present polemical papers for public consumption, but in the workshops, Heidrun found that most participants were striving to achieve an even-handed understanding of the "imperialist" powers.

Nevertheless, she found herself being pulled back into the obscurantism that she had briefly escaped from during her sojourn in the United States. It seemed increasingly ludicrous to try to make sense of the United States through the Communist Party press. Special permission was required to read *US News and World Report* or *Time*. Relevant books were equally difficult to gain access to. When Heidrun tried to obtain titles by Herbert Marcuse, she had to request them from East Berlin. The Berlin authorities told her that she would have to have special authorization to read Marcuse. As Heidrun points out, "It didn't make any difference that he was a Marxist. In any case, they had never read Marcuse, so what did they know about him!" Marcuse and many other writers were on the "poison shelf" (*Giftbücherbrett*).[1]

As for teaching, though, Heidrun had considerable freedom within her department to teach about America as she saw fit. When she talked about the American Constitution, she held it up as the model for democracy that East Germany should draw on in its attempt to develop a democratic socialism. She shared a feeling with other members of her department concerning "lost opportunities," which, if taken, would have put them closer to a perfected socialism as the next step in the evolutionary development of mankind. The lost opportunities Heidrun and her colleagues were referring to were the systems of democratic process that already existed in American society but were rejected by party authorities as bourgeois praxis.

For Heidrun, the disparity she felt between the lack of democratic procedures in her own situation and their existence in the West led to dismay. Yet in the context of the Cold War, Heidrun says, "I belonged to the East. . . . I still believed at that time, in the early eighties, that basically the Russians were more right than the Americans—that the impetus for the Cold War came from America. I believed that the Russians were acting in their own defense, even though neurotically with their overemphasis on military security. It was only when I began to realize that neither the Americans as individuals nor the Russians nor we could exert any influence on the military—that we were all entrapped—that I slowly gained a more sophisticated perspective of the Cold War. But that emerged only after my second trip to the United States. I suppose you can say that I was just a Cold War kid and knew basically only the one side."

The Second Invitation

Heidrun's second invitation to the United States came as unexpectedly as the first. When the Department of American and English Studies at Humboldt University in East Berlin entered into an affiliation with a large midwestern university, the department transferred the relationship it had had with Colby College in Waterville, Maine, to the University of Jena. Professor John Reynolds of the German department at Colby College lost no time in organizing an invitation through the Ministry of Education of the German Democratic Republic (GDR). The list of candidates was headed by a colleague of Heidrun's who had received her degree in 1981, a year before Heidrun. When she married and became pregnant, the invitation to teach literature at Colby for a semester passed to Heidrun. Heidrun had never taught a literature course, German or otherwise, but that did not deter Colby, eager to have someone on its staff from exotic East Germany, nor did it trouble the ministry, pleased to have the representation of GDR culture undertaken by a teacher of proven socialist sensibilities. Since the diplomatic recognition of the GDR in 1973 by the United Nations, there had been an increasing interest in East German literature in the West. Professor Reynolds was able to assure Heidrun that she would find Christa Wolf on the college's shelves, both in the original language and in translation. Heidrun says with a laugh, "There were some GDR books which I had not been able to read here, which I read there!" It was common knowledge, according to Heidrun, among those who taught East German literature in the United States that the most authentic information about the GDR came out of the country through the people who taught those courses.

The United States Revisited

In early February 1983, Heidrun left for her second visit to the United States. This time, she was to fly business class on Lufthansa from Tegel Airport to Boston, expenses paid by Colby College. After she passed through the draconian controls at Friedrichstrasse, she continued by subway and bus into West Berlin for the first time.

Eager to be returning to the United States, her arrival in Waterville, Maine, on a cold, wet, February day deflated her spirits. Edgar Schick, now president of a small college in Sanford, Maine, met her at the Portland airport the night before, after a connecting flight from Boston. Reynolds picked her up at the Schicks' house the next day and drove her to Waterville. The room she was brought to in a student dormitory was bereft of everything except for a bed, a bureau, and a chair. She found her way to the dining hall and asked for a cup of coffee. No one contacted her until the next day. For the first time in the United States, she felt a pang of homesickness, although in time she began to feel more at home as colleagues, step by step, provided her with the things she needed in her student dormitory surroundings.

Even before leaving Germany, Heidrun knew that this second American experience was going to be different from the first. Her short trip in 1979 was a visit, and she, a visitor, looked at things from the outside. She had spent most of her time in libraries. Now she felt that she was living in America, rooming in a dormitory with students, sharing experiences with colleagues. They invited her to their homes. They went out on the town together. She was able to compare a depressed Maine lumber town, Waterville, with the more affluent high-tech ambiance of Rochester. Daily life in America became Heidrun's reality.

What struck her most was the difference in the nature of relationships between people in America and people in Germany. People everywhere seemed friendly, open, hospitable, easy to meet, and generous. As Heidrun says, "If you call somebody in Germany on a Sunday and say you are so-and-so from America, they will think, 'Oh my God, how terrible, what will we do?' But in the United States, they would say, 'Why, of course, we will meet you! Stay there, we will be right down to pick you up.' But on the other hand, friendships don't seem to be as deep in America as they are in Germany. When things become serious or difficult, people withdraw into their family, into themselves, and appear to lose interest. That, of course, is a generalization, and there are exceptions. I think I probably know more exceptions. I think that it may have to do with the fact that in America, everyone is on the move, and so are relationships. I remember one friend saying to me, 'I have stopped seeing her, but we are still good friends.' That was hard for me to understand." It also took time for Heidrun to become accustomed to calling everyone, colleagues and students alike, by their first names—most particularly the department head. The first time she met the chairperson of the department, she addressed him as Dr. Weiss, but after that it had to be "John."

Since Heidrun had attended a church function with the Schicks in Rochester, she had been impressed with the community-building role that churches assumed, and the ways in which parishioners participated in assuming responsibilities for church affairs: running bake sales, organizing drives for blood donorship, or manning a soup kitchen. She even found people who looked around to join a church of their choosing rather than just belonging to one from birth, as was the case in Germany.

She had her first experience, too, with New England town meetings, at which the members decided what the community tax for the coming year would be. Heidrun says, "I had been a member of the town council in Göttern for ten years, and we never had a chance to decide how much money we needed to spend on something. The money was allocated to us by others. In local democracy, I saw this kind of community working for the first time." These and other examples of participatory democracy—voting for judges in local elections, for example—called into question the picture she had received at home of oligarchic rule imposed by American monopoly capitalism. At the same time, Heidrun found no contradiction between her admiration for the American democratic process and her continued belief that America was an imperialist power, bent on exerting its dominance over other nations by threat and force.

Heidrun had more time to travel on her second trip. She visited friends in Boston, and from there went to Rochester, then to Hanover, Indiana. Heidrun says, "My friends from Madison had moved to Hanover, so I spent some time with them in Indiana. That was interesting, because they took me to Ripley, Ohio, knowing of my interest in the Underground Railroad. One of the agents had a house up there on the banks of the Ohio. You got a really moving feeling for what it must have been like to cross the Ohio and to be taken care of and hidden. It was a very important thing for me. History was becoming real; all that I had read and taught to my students—you could see it right there."

In regard to feminism, Heidrun initially believed that she had little to learn from America. "I thought," she explains, "that we were ahead of the rest of world. At Colby, students and colleagues asked me about kindergartens, day care centers, 'mother years'—all those practical things that we had in place for such a long time. It was accepted in East Germany that women were not only housewives but actors in the national economy. Women were not displaced in East Germany after the war as they were in America. There was no need there to develop the 'feminine mystique.' That women worked was taken as a matter of course and of necessity. We had our own feminist writers—Helga Königsdorf, Brigitte Reimann, and Christa Wolf, in her own way. When I came to the States, I thought I was a pretty liberated woman, which I was. Everything had worked out nicely with my personal life, with my family, with having a child and a career. I had not come to any limits where I pushed and felt 'there is no way; you can't go any further than that.' Having just finished my dissertation, I was still at the stage where all the men patted you and said, 'Well done, girl,' and they didn't see the potential competitor in me yet. That came later. Then you realized that men were very alert when it came to prestige, and if there was a woman who was trying to be on par with them. . . ."

Heidrun attended the weekly feminist lunch group at Colby and went to a number of their demonstrations, events that did not occur in East Germany, where a woman's role in civic life was guaranteed by the constitution.[2] What women often talked about at Colby was something Heidrun had rarely heard mentioned at home: male attitudes toward women. As the analysis developed, Heidrun realized that the attitudes identified by her feminist Colby colleagues were similar to those that prevailed among her male colleagues at the University of Jena and were merely taken for granted. The conversations brought to mind an incident that had happened in her department in Jena when a replacement was being sought for a colleague. Heidrun says, "I said, 'How about Doris Schönfeld as a replacement?' and it was as if I had not said anything. Later another woman said, 'What about Doris? Isn't she qualified?' Nothing . . . until eventually one of the older men said, 'Well, maybe we should consider Doris Schönfeld.' So I thought, 'Hey, what's going on here?' As a woman in such a group, you can make very sensible suggestions, and they just don't listen to you—they don't hear it." Her own unawareness of these behaviors made Heidrun recognize how pervasive male oppression of women in the GDR was. Heidrun began to feel her consciousness raised in a way she had not foreseen.

Heidrun "on the Circuit"

Heidrun returned to Germany in June 1983. She was met at the East Berlin airport by Erhard and Maria. While waiting to be cleared in the customs area, Heidrun could see Maria's slight form bent over as she tried to peer at her mother under the barrier. An empathetic customs officer, watching Maria, let her through the barrier. She ran into her mother's arms.

Instead of remaining overnight in Berlin, they came immediately back to Göttern, as Heidrun was looking forward to seeing everyone again. She had so much to talk about. Yet she found that as she tried to answer the family's rapid-fire questions, no one was really listening. It was Erhard with whom she could speak at length about the trip.

One of the first things she did was to work up two lectures—one on Washington, D.C., and another about a trip along the East Coast—for the Urania society, which booked speakers for schools and work collectives for a modest fee of 30 deutsche marks (DM) per lecture. In her presentation, Heidrun was careful to stay away from sensitive areas that might suggest invidious distinctions between East Germany and the United States. She found the older audiences interested in what she had to say, asking her questions about how Americans lived. Students, however, were invariably disappointed that she did not have more information about American pop culture. She was "on the circuit" for almost five years. As Heidrun says, "I certainly didn't do it for the money, but rather as my way of sharing with others who would never have the opportunity of visiting the United States."

Heidrun and the Peace Movement

When Heidrun returned from America in the late spring of 1983, her four-year contract at the University of Jena had expired. She was given a one-year contract, at the end of which an opening in the department became available, and she was awarded tenure. Tenure was granted to anyone who had been an assistant for four years, who demonstrated promise as a teacher and researcher, and for whom there was an opening. The early 1980s were still a period of expansion and growth at the university. For Heidrun, tenure meant a little more security and time to do different things, such as write about the American peace movement, to which she had been introduced while at Colby. As it turned out, writing about the movement had a determining influence on a number of her opinions.

Soon after her return to Germany, a conference on American studies was held in Jena, to which a number of American scholars were invited. At that time, Heidrun was still of the opinion that the GDR was with Russia on the side of peace. She remembers having a lively discussion with Michael

Geisler, then of MIT.[3] Geisler argued for unilateral disarmament, affirming that "the ice had to be broken" and that someone had to take the first step. Heidrun disagreed, insisting on the right of the Soviet Union to retain its missile capability for defensive purposes.

Jena itself had become the center of the unofficial peace movement, centered on the group Swords to Plowshares (*Schwerter zu Pflugscharen*). According to Heidrun, the name was derived from the statue that the Soviet Union gave to the United Nations in New York, depicting a powerful man transforming swords to plowshares. The position of the group was that the Soviet Union should begin to disarm immediately, starting with its installations in East Germany. Heidrun was aware of the Swords to Plowshares movement but viewed it with ambivalence, fearful that it might covertly support Western interests yet feeling persuaded by the integrity of its conviction.

Heidrun remembers going to an official demonstration sanctioned by the party in Jena in 1985. She could not help but notice a side demonstration against atomic weapons being pushed out of sight of the platform by plainclothes officers. She could see the demonstrators' slogans, and they said nothing that she disagreed with: "I was really moved, and I thought, 'Why are they trying to push these people away?' But I wasn't brave enough to say, 'Hey, I am with you.' I was still so much in my old way of thinking. I thought, 'Well, they are right—there are many bad things, but basically we are on the right track.' So I didn't know what to make of it. I walked around Jena all day after that, and I didn't know what to do. I thought I should be with that group of people, but on the other hand, I still belonged to the other side somehow. I had an obligation, a feeling of where you come from, of where you belong." It was the first time that Heidrun had allowed herself to be so moved by the appeals of the peace movement that she found herself agreeing with them, but what her mind affirmed, her heart refuted. She valued too highly the ties with her comrades to break rank.

What did begin to make a difference in Heidrun's thinking about the Cold War were the writings and activities of the Berrigan brothers in the United States, about whom she had become informed in the course of undertaking research on the peace movement in Britain and the United States. Their insistence that further superpower confrontation would lead only to disaster, that old ways of thinking had to be discarded, and that the shield must be laid down and one's trust placed in a love of humanity rang true for her as no other arguments had.

An equally powerful influence was Henry David Thoreau's defense of the individual's conscience in his *Civil Disobedience*, as well as the American Quakers' strong indictment of war and commitment to the alleviation of international tension. As Heidrun says, "Somehow the American scene was closer to me than the German one. I lived here in an 'ivory tower.' I did not see or hear much of what happened around me. I was much more connected with things that came from the United States, and I felt that if Americans were talking this way and if we could join with them somehow, then we could lessen the tension. I didn't have these discussions with students at Jena but rather with Americans who came and commented on the situation."

Heidrun's increasing interest in the peace movement and the nonviolent tradition on which it was based carried a price. Occasions to which she had always had *entrée* were now barred to her. Soon after her return from the United States, she had planned to attend a function at the American embassy in East Berlin with her mentor, Professor Leitel, but she received notice from the Department of International Relations that she did not have permission to go. She was also informed that she could no longer apply for grants.

Nevertheless, the momentum of change in Heidrun's thinking that occurred in 1984 and 1985 following her return from her second trip to the United States was further stimulated in 1986 by ideas of socialist reform then emanating from the Soviet Union and the 27th Party Congress, in which Mikhail Gorbachev began to talk of the need for "new thinking."[4] Gorbachev's perestroika came as a revelation. A sense of euphoria took hold with the recognition that Gorbachev's ideas were the right way. "'That's what should happen here,' is what we thought," says Heidrun. "Now, there were new ideas and the belief that somehow we would be able to change socialism for the better."

Much of that heartfelt conviction was based on the belief that the GDR, compared to other Eastern European countries, was strong enough economically to survive the restructuring that reform would entail. East Germany would not collapse. In retrospect, perestroika was, Heidrun admits, "a tremendous self-deception. Yet there was so much hope."

The Party School

In 1986, the time had come for Heidrun to spend a year in a party school, an obligation required of everyone at some point in their careers. It was not something that Heidrun looked forward to. She was scheduled to spend three months in a school in Bad Blankenburg, living in a dormitory and attending seminars on Marxism and Leninism. The thought of the dormitory repelled her. Fortunately, there was a way out. Designed for mothers, the Women's Special Class (*Frauen-Sonderklasse*) allowed her to live at home, attend classes in Jena, keep in touch with the university, and supervise her students' theses.

The eight-hour day was spent taking classes, attending seminars, writing papers, and taking exams. Heidrun's group was made up of women from different walks of life—professional persons, academics, workers from the Zeiss factory. The experience turned out to be far more worthwhile than Heidrun had expected it to be. She says, "We also had time to study the classics—to read [Karl] Marx and [Friedrich] Engels again after such a long time. To be able to go back to the roots, so to speak, and at the same time to have the new input from Gorbachev, gave me the feeling that socialism could work if we did it the right way."

Part of the excitement of the year lay in the SED's inability to prevent a discussion of the Soviet Union documents promulgating reforms. It was an expectation of the school curriculum that the party would reply to the comments sent to it by the seminar participants on the basis of their reading, discussion, and reflection. In 1986, no comments were forthcoming. As Heidrun says, "They couldn't say no, and they couldn't say yes. It was as if they were paralyzed. But that was good, for if they had not been paralyzed, they would have sent the army out."

The excitement and sense of hope generated at the school was carried back to the department and the classroom by Heidrun and the others. Intense discussions were held in 1987 and 1988 about how the course of study could be restructured and the ways of doing things reorganized. Some changes were made in course content and in modes of teaching. For the first time, Heidrun believed that teaching the American Constitution was an entirely logical and natural thing to do. Yet as 1988 gave away to 1989, the feeling that nothing was going to happen was pervasive. It was a common experience that no matter what was suggested, nothing came back. Everything stayed the way it was. There was not even repression coming down from on high that could be struggled against. Heidrun found herself withdrawing again into a world of ideas. She says, "That's the good old German way. In my case, I studied American ideals to distract myself from German reality."

The Unthinkable Happens

In the summer of 1989, Heidrun had the opportunity to travel to England for a three-week course in language teaching. She was eager to go, to get away. The feeling of hope that had been so strong in 1986 had given way to despondency. Heidrun says, "For me it was depressing, because I felt that the old ways did not work any longer. I did not see any place where I belonged. I was not linked to any of the new groups, like Neues Forum.[5] I didn't know how to talk to them, and they probably didn't know how to talk to me. At the same time, I no longer belonged to the old Leftists. I was just alone, thinking, 'Oh, God!' I was still talking to my friends, but we had stopping saying anything."

The trip to England was a distraction. Heidrun came home a little more relaxed. Yet it was a sad time. Heidrun found herself going through the motions of teaching. On the evening of the November 8, she attended a Göttern council meeting and went to bed immediately on coming home. She was up at 4:45 the next morning to catch the bus to Weimar and the six o'clock train to Berlin to attend a meeting at the Ministry of Education for IREX (International Research and Exchanges Board) scholarship applicants. She walked to the ministry from the train station and noted how empty the streets were for a Friday morning. She says, "I knew something was up when I went to buy bread and there was no line." At the meeting, she met a friend from Jena. She sat down next to her. Her friend said, "Oh, I am so excited. I was up all night watching TV; weren't you?" When Heidrun asked her what had happened, her friend replied, "Don't you know? The Wall is down!" Heidrun says, "I couldn't believe it! Things happened so suddenly. But I wasn't sad; rather, I thought, 'Now this is the way things should be.' All the same, I couldn't help wondering what shape our government was in, that something like this had happened. The only thing that we can give credit to our leaders for is that they had enough good sense to tell the soldiers not to shoot at the people. But even a great thing like the opening of the Wall happened not because of the government's decisive leadership; it appears to have happened because of default. In that respect, it was rather disappointing. From the other point of view—that people just pushed, little by little, and got what they wanted—in that respect, it was great!"

The optimism that perestroika had provoked in 1986 but that had all but disappeared by 1988, Heidrun now found rekindled in herself and among her friends. Heidrun says, "I thought [the Wall coming down] was a tremendous opportunity to make the changes that we wanted to see. I did not want to abolish the GDR. I grew up in this country—it was my country. I had no emotional relationship with West Germany, and I thought that this was a necessary step to improve things."

A Quick Look at the West

Heidrun had already been in the West, albeit briefly, and Erhard and Maria had crossed the border on a very cold day in November to go to Bavaria with friends, just to see it. Now, before Christmas, a family trip was planned, "not to celebrate our new freedom but to get some money that the State of Bavaria and the city of Munich decided to give to their poor cousins in the East," Heidrun says. "Of course, their poor cousins in the East were happy to get it." When Erhard came home one night and said that two of his colleagues had taken the free train from Berlin to Munich and found Munich very nice, Maria would hear of nothing other than Munich as their destination in the West. Heidrun attempted to get reserved seats, but at the station ticket counter, she was laughed at. "Standing room only!" she was told. Heidrun shuddered at the thought. She disliked being in crowded places.

It was finally decided that they would drive the Trabi across the border to the nearest town and take a West German train to Munich. Erhard had already bought the tickets, Saalfeld to Munich, paying for them in ostmarks. They got off to a very early start on a Saturday morning a week before Christmas. Heidrun remembers the excitement she felt as they approached the small crossing in the woods, the Trabi swaying and lurching as it crept over the bumpy road: "It was a great moment—just going out together with them. It was a very moving experience."

The wait to receive the money in Munich was long. Heidrun felt defensive about her compatriots standing in line. Somehow their appearance offended her, and she wondered whether the West Germans felt that way, too. Erhard, Heidrun, and Maria each received 90 DM. It was the most money Maria had ever had. She invited her mother and father to McDonald's. "Now that I have money, I can invite you," she said. Afterward, they visited department stores and just looked. Before heading back to the East, Maria bought herself a record, *New Kids on the Block*.

Crossing the border so easily had been exhilarating. Heidrun knew from her own experiences in the Soviet Union and in the United States how important travel could be in informing one's opinion. She thought that now that people could freely travel, they could shape their own opinions rather than just follow the party line.

A Time of Infinite Possibilities

Heidrun and her colleagues again felt empowered. The SED rapidly disintegrated. Now with Big Brother no longer looking over their shoulders, they had the space to try new things, to introduce a new system. In conjunction with others, Heidrun's department set about reorganizing the university, creating a structure representative of the staff and the faculty. Support was forthcoming from professors from the West, who had been involved in the student movement of 1968. They told them to be careful not to let the same thing happen that had happened to them—allowing their achievements to be taken away. As Heidrun says, "We assured them that we would not make the same mistake."

It was a time of seemingly infinite possibilities, with no end of interesting topics to work on. Student grants became available. Students in Heidrun's department were sent to England, some to the United States. The Kennedy Library in West Berlin was available for research. Guest speakers came, one after the other. The department head was elected, not just by professors but also by assistants and students. Departmental outings were a frequent event. The book Heidrun co-authored, *Peace Movements in Britain and the USA*, was published in 1989, and two papers, one on the American peace movement and the other on Frederick Douglass's advocacy of women's rights, appeared in scholarly journals in 1990.[6] Heidrun looks back on 1989 and 1990 as "some of the best years of my life."

Toward the end of the year, the whole department went to England for a week of seminars at Jena's partner university, Canterbury, with expenses paid by the British Council. There was a feeling even then, though, that Canterbury was the last great outing and that before long, people would be let go from the Jena department.

The demise of the GDR and the SED also meant the end of the Department of American and English Studies, as it had been originally constituted as an institute of research on imperialism. The election of March 18, 1990, which brought the Christian Democratic Union (CDU) to power, signaled the beginning of the restructuring of the department from the top. Heidrun knew that sooner or later, she would have to leave. At the same time, many were abandoning the SED and joining the Party of Democratic Socialism (PDS). Some did it quickly, surrendering their red membership booklets; others did so slowly. Heidrun was one of the last to do it. She says, "For me, it was not that easy. I had joined the party for idealistic reasons. I told myself there were party functionaries who had betrayed the idea, but somehow *you* have to stick with it. So I stayed. I think I was one of the last people in my department to hand in their documents.

"I attended meetings of the PDS and those of some other Leftist groups. There, I recognized the same faces. People were trying to look for a new home where they thought they could belong, somewhere that was Leftist but not in the old ways, a new Leftism, but it didn't materialize. Once you identified yourself as a member of the PDS—the successor of the SED, so to speak—you were out; you were not taken seriously in any discussion. The only other thing you could do was to sacrifice yourself—become a martyr to the cause—so I left."

Always a strong believer in the party, Heidrun now found political parties and organizations to be the principal obstacle to the idea of a socialism informed by the tradition of American participatory democracy. For that reason, she stopped attending the local meetings of the German women's organization (the DFD).

"It seemed impotent at that moment," says Heidrun of the political situation in the newly unified Germany. Disenchanted with political parties but not with politics, she retained her seat in the Göttern Village Council, an institution she esteems for its commitment to the democratic process.

In December 1991, Heidrun left the university, convinced that she would soon be let go. New professors and their assistants were being brought in from the West, most of the former faculty having been found inadequate. Heidrun says, "It was a mistake. I should have hung on like some of my colleagues." Heidrun lost her status and benefits in civil service by leaving so precipitously. She was prompted in her decision by an offer from the Elter School in Weimar to teach English to women being retrained for positions in the new market economy. She was given a week to make up her mind. Confronted with the possibility of being without work—a prospect she had never faced before—she accepted the position. Heidrun attributes what she now considers "my big mistake" to her inexperience in having to make career decisions. A premonition of the uncertainty that lay ahead for her and her family as a result of the freedoms bestowed on them by the West began to overcome her.

During the winter of 1994, the Elter School experienced difficulties recruiting students, jeopardizing Heidrun's position. It seemed that her worst fears were being realized. She began to look for employment elsewhere. All teaching positions in public institutions required applicants to have a civil-service rating. Only jobs in private schools or in foreign countries were available to someone in Heidrun's position. She applied for an opening to teach German in a high school in Alabama but was turned down. By the time her place at the Elter School was terminated at the end of March, Heidrun had begun to think that she would have to leave teaching for some other field. She considered the possibility of going into business with Erhard, promoting Weimar as *Kulturhauptstadt, Europas, 1999* ("European Capital of Culture 1999"); he would do the visuals, she the text.

The spring of 1994 was proving to be the most troublesome of Heidrun's life. Everywhere in print were discussions of the severe problems women of the former GDR faced in the new united Germany. Erhard and Heidrun were thankful that they and Maria had a place of their own to live.

At the end of April, Heidrun applied for an opening teaching German in the Lithuanian city of Panevėžys. To her delight and considerable relief, she was accepted. Her salary would be paid for in part by the Lithuanian government, in part by the German. The take-home pay would be the most she had ever received. Heidrun was assured that Erhard and Maria would take care of each other in her absence. Plans were made for them as well as her mother to visit her in Panevėžys. She worried, though, about her father and his failing health, now that she was to be so far away.

In mid-August 1994, when she drove off from Göttern in her secondhand, heavily laden Citroen, she believed, as she later wrote in a letter, that she was embarking on a new phase of her life.

Commentary

Eastern German women constitute about 10 percent of the total German population.[7] Since unification, they have been the subject of interest of a wealth of former East German social scientists and commentators as well as their West German counterparts.[8] Much of the discussion has been focused on the loss of economic independence of these women and the prospect of their being pushed back into the role of housewife.[9] The fact that employment is central to Heidrun's own sense of self points us back to the GDR for clarification.

It was a widely held opinion, most especially in the East, that the position of GDR women was more advanced than that of German women in the West. The granting of equal pay for equal work, the protection of married and single working mothers, the right to abortion on demand for women in the first trimester of pregnancy, the awarding of maternity leave of a full year off with full pay, free prenatal and postnatal care, and free childcare—all were enjoyed by women in the East. As early as 1971, the GDR leaders proclaimed that gender equality had been achieved.[10] By 1984, approximately 80 percent of GDR women of working age (between eighteen and sixty) were employed.[11] In 1988, 82 percent of working GDR women had completed some form of job training.[12]

The realities of everyday life, though, revealed discrepancies between what was being proclaimed and what women actually experienced. This was apparent in, among other places, the gender-specific nature of their employment. Although strongly overrepresented in the service industries, they were underrepresented in industry, construction, transportation, and management positions. Women consistently occupied the lowest-paying jobs.[13] In addition to being fully employed outside the home, women were responsible for the greater part of the housework, as well as all those small tasks and labors of love that ensure familial well-being. As Hildegard Maria Nickel points out, the stress of this "double burden" (*Doppelbelastung*) made it difficult for women to fully realize the cost of the equality that had been proclaimed in their name.[14] Relief from the strain of overemployment was sought among some through abortion and divorce.[15]

Nevertheless, despite the downside of the position of women in the GDR, their self-perception as essential working members of society was deeply rooted. Not overlooking the fact that work was a duty as well as a right, it was, as Nickel affirms, part of GDR women's identity to be gainfully employed, enjoy legal autonomy, and gain some degree of financial independence.[16]

In West Germany, employment for the great majority of women was not a national expectation, nor did the institutional infrastructure for its implementation exist. West German maternity leave of six weeks before the birth and eight weeks after stands in sharp contrast to the time granted the mother in the GDR.[17] The extensive provision of childcare facilities in the East was not equaled in the West, where mothers often had to make alternative arrangements with neighbors, relatives, or friends. As Mary Fulbrook writes, "While East German women took it for granted that they would return to work, West German women found considerable obstacles and barriers to be surmounted before they could consider doing so. . . . [F]or West German mothers, not working was the norm."[18]

It is clear that the importance Heidrun attaches to gainful employment stems from an Eastern socialist tradition of work as right and duty rather than a Western capitalist perception of work as economic necessity. Even though doing so would have been a hardship, Heidrun could have remained at home with her family. That she chose instead to assume a position in Lithuania is a reflection of the relationship she shares with women of the former GDR regarding work, economic independence, and personhood. The price that others pay for failing to find employment in a market economy is to become, as Ursula Schröter points out, "losers" in their own self-estimation.[19]

The satisfaction that Heidrun is experiencing as she begins her second year of teaching in Panevėžys is undoubtedly due, in no small measure, to her own temperamental inclination to persevere and to excel. On more than one occasion, she has admitted, with a mixture of pride and self-deprecation, how far she has come from her origins, a daughter of "peasants," while at the same time giving credit to a regime that made it possible for her, a child of the working class, to attend the university. She counts the blessings that came her way as a student at Jena, her doctorate, her colloquial knowledge of Russian and English, and the ways those skills allowed her to feel comfortable in the East and in the West. Heidrun hopes to be able to return to the United States and to teach again, if not at Colby then at a similar liberal arts college or possibly a university. She says, "I feel I am ready to think further about some of the feminist issues that my colleagues at Colby helped me with when I was there in 1983." Heidrun has in mind her own developing sense of autonomy, a feeling about her selfhood that she recognizes as being informed by her "American experience, the constitutional guarantee of individual rights." More forceful, though, is her awareness of her indebtedness to the GDR "for providing me with the tools and the empowerment to become a competent person in my own right."[20]

PART IV

19

Maria Schorcht

School

MARIA SCHORCHT, the second oldest of Edgar and Wally's six grand-children, turned seventeen on January 4, 1995. Maria lives on the second floor of the Schorcht house in Göttern, just down the hall from her parents and grandparents. Maria also remembers playing as a child in the barnyard and beyond the gate, in the village. She especially enjoyed meandering along the banks of the Magdel, hiking in the woods, and climbing trees. Now, except for sleeping, she is rarely in Göttern on weekdays.

On school days, Maria struggles out of bed at 5:45 A.M., showers, eats breakfast, and leaves the house at 6:30, on the run to catch a ride with a friend to the Friedrich-Schiller-Gymnasium on Thomas Mann Street in Weimar. Maria transferred from the Goethe High School to the Schiller because of its wider range of course offerings. The Goethe school, humanistic in its orientation, is well suited for the student wishing to major in German, history, or English. The Schiller school, scientific in its emphasis, corresponds to Maria's newfound interest in math and biology. At first, she had trouble keeping up with calculus—the classes were more advanced than any she had taken before. Thankfully, however, she could take a remedial course. About mathematics, Maria says, "It teaches you to reason and to be logical. You can't just pretend to know something, as you can in the 'wordier' classes, like German or English. In my literature courses, they always asked me to find things in the author's writing that I never believed he actually intended to be there. And even if she or he did, it wouldn't matter, because it is the reader who counts, not the author; he had already done his work."

As for biology, Maria points out that there is not too much theory but a great deal to learn by heart. There are also laboratory reports to prepare. Maria says that at first she was flustered by the long papers that some of her classmates wrote, but when she found that they were mostly filled out with cover sheets, tables of contents, and appendices, she felt better. Maria says, "I have a hard time writing five pages, the way some people do, to say nothing of twelve."

In 1984, when she was six, Maria entered the first grade in the Felix Dzerzhinsky polytechnical school in Magdala, the same school her father attended twenty-eight years before. She and her cousin Katherina, her friend Yvonne, and their friend Thomas took the bus every morning from Göttern for the short ride to Magdala, arriving just before school started at 8:00.

From grades one through four, Maria was a Young Pioneer (*Jungpionier*). She wore a white shirt with the insignia "JP" sewn on it and a blue kerchief. Then in the fifth grade, she advanced with her classmates to become an *Ernst-Thälmannpionier*, substituting a red kerchief for the blue one. It was a big step. Assemblies were held on special occasions, such as Ernst Thälmann's birthday or May 1, when awards were handed out for a poetry reading or a musical composition. On those days, all the Young Pioneers were required to repeat the Thälmann oath, pledging to be always prepared for "peace and socialism." It was also expected that each Pioneer group would have a brigade partner in the community, so that children and adults might share a sense of socialist solidarity. For Maria's group, their brigade partner was *Die Ambulanz*, the staff of the Magdala Polyclinic. On certain days of the year, the two groups would perform for each other—songs and poems by the students, and rescue drills by the staff.

Pioneers were supposed to carry their Pioneer ID cards, but Maria often had difficulty finding hers at the precise moment when it was most required. On it was a passport photo, date of birth, signature, and space for the proper authority to write in the position held, whether council member, deputy, secretary, or agitator. Maria was an agitator, meaning it was her responsibility to inform her classmates of news items they should be familiar with. Maria chose not to comment on political topics covered in the newspapers. "Politics are too complicated," Maria says. She chose instead to talk about discoveries made by archaeologists and other scholars. "I'm pretty sure, though," she says, "that is not what the agitator was supposed to be doing. Actually, I had no idea what I was supposed to be doing."

Maria remembers the time they prepared a bulletin-board display for the whole school. It featured a picture of an East German tank with a caption reading, "Our army securing the peace for our country and its borders," to which Maria and her fellow Pioneers added, "We don't like tanks or wars." This sentence was cut out before the picture was mounted. When asked why, the teacher replied, "Because the caption was too long and did not go with the picture." Maria says, "That is when I noticed that something was wrong. It must have been in the sixth grade."

By the time Maria was old enough to join the Free German Youth (FDJ), the German Democratic Republic (GDR) educational system had come to an end with unification. The traditional school system in the West was reintroduced into the East in 1990–1991. The comprehensive and democratizing polytechnical secondary school was transformed into a *Regelschule*. Between them, Maria and her father had frequented the Magdala polytechnical school for almost the entire span of its forty years of existence. Erhard entered in 1958, six years after the school's opening. Maria left in the spring of 1991, the year it became Westernized. In September of that year, she entered the Goethe *Gymnasium* to obtain her *Abitur*.

Looking back now from the vantage point of the Goethe and Schiller schools, Maria regrets that some aspects of the GDR educational system do not still exist. One of these is the student-pairing system. She says, "I was paired up with poor students because I was good, and I helped them with their homework after school. But that program doesn't exist anymore. Now if you are having trouble, you have to go find help somewhere on your own, or you don't get helped." Maria liked the idea of assuming responsibility for each other. Presently, it seems to her that people are looking out for only themselves. The concept of helping others, she explains, used to carry over into the community as a whole. Each student was assigned an elderly or an infirm person to look out for, helping with the shopping or accompanying this person on errands. Maria says, "To help others and to be a good friend was written into the student statutes. You have to struggle to get people to do that now."

Maria knows several former classmates who remained in what is at present the *Realschule*, or *Regelschule* (intermediate school) in Magdala. Now when she meets them, she senses an edge of enmity in the relationship. She says, "People in the *Regelschule* don't like those of us who went off to the *Gymnasium* to get the *Abitur*—they have the feeling that we look down on them. It is new that the good students should leave and go to another school, while the bad ones stay in the *Regelschule*. Here another selection takes place, so that the least-gifted students end up in the *Hauptschulklassen*. That creates division among us. There is a lot of stuff going on in those schools, violent things. It was not like that before, when we were all in one school."

Domesticity

If Maria has school until 2:00 or 3:00 in the afternoon, she visits with a friend in Weimar and then goes home with Roswitha at 4:00. If she gets out earlier or has to stay later, Erhard makes arrangements to pick her up. If all else fails, she can catch the bus, which takes an hour, costs 4 marks, and does not get her home until 6:00 or 7:00. Depending on the time of day, she prepares something for her father and herself for lunch or dinner or for both meals. After cleaning up, Maria plays the piano to relax, a classical piece or jazz or boogie, depending on her mood. At 8:00, she starts her homework while listening to music. On the weekends, usually Saturday night, she goes out and does not return until the morning hours.

Since her mother received a job teaching German in Lithuania, Maria has been doing the cooking for her father and herself. She misses her mother but rarely talks about it. She realizes that to do so would only fuel Edgar and Wally's resentment toward Heidrun for defaulting, in their opinion, on her responsibility as a mother. As Heidrun acknowledges, they are worried about Maria staying overnight with her friend Yvonne in Weimar, not telling them about her plans, and so on. Maria has become quite independent in her ways, according to Edgar and Wally.

Maria likes to cook for her father because it makes her feel closer to him and to her mother and her grandmother as well. Maria and Wally share an affinity for baking. Growing up, Maria learned of its importance in the family. When she was twelve, she wanted to turn the *gute Stube* into a bakery shop, with its window facing conveniently onto the sidewalk. Wally was delighted with her granddaughter's plan, although she knew it was not to Edgar's liking. Even though the project came to naught, Wally did not relinquish the hope that Maria would one day take up baking seriously. It would be her legacy.

America

One theme running through Maria's life that she does not share with others her age is her privileged relationship to America. As a child, she remembers her mother talking about her first visit to America and then living and working there for six months. Maria was both proud and envious of her— proud that her mother had been to that fabled land, yet envious that she could not have gone also. Heidrun encouraged Maria to learn English at school, helping her in every way with its rudiments. By the time she was twelve, she had begun to believe that those American places, documents, and events that meant so much to her mother—New York, Boston, Concord, Seneca Falls, the Emancipation Proclamation, the Underground Railroad— had become part of her own life.

In the summer of 1991, the opportunity came to see them for herself. Heidrun wanted Maria to see America with her own eyes, so she and Erhard planned a family vacation to the American East Coast. Through his journalistic contacts, Erhard managed to obtain a good price for the flight.

Heidrun had friends along the way with whom they could stay. Today, much of that trip is still vivid in Maria's mind: the excitement of New York City, where Erhard made a documentary video; the Freedom Trail in Boston, along which they retraced the first steps of the American Revolution; Concord, where they walked over the "rude bridge that arched the flood";[1] and their pilgrimage to Seneca Falls, the sanctuary of the women's movement. Everywhere they went, Heidrun, drawing on the lectures in American studies she had given at the University of Jena, enriched the place by sharing stories of its history.

When they returned to Göttern after three weeks of U.S. travel, everyone was glad to have them safely home again. Maria found, though, that while they all said they wanted to hear about the trip in detail, she was not able to truly communicate with anyone about what the visit had meant to her. What was certain, though, was that she had already made a place for America within herself. She decided that as soon as she was old enough, she would go there as an exchange student.

The opportunity to revisit America came earlier than Maria had expected. In return for the two tours that the high school choir of Statesboro, Georgia, had done in Weimar, the fifty-voice choir of the Goethe High School made plans to sing in Georgia in the summer of 1994. Maria looked forward to the trip, even though she intuited that it would not be the same as the one she had taken with her parents. As it turned out, she was right. Whether due to the absence of her mother's informed appreciation, which suffused everything she saw with a kind of wonderment, or whether at age sixteen she had developed a more critical, discriminating way of observing things is difficult to say.

Following an initial week in Statesboro, Georgia, the choir went on a tour through the southern part of the state to Birmingham, Alabama, where they remained for another week before flying home from Atlanta. Their repertoire included old German masters from the fifteenth century, German folk songs—especially those from Thuringia—some spirituals, Beatles songs, and Bach, intermixed with instrumental pieces.

There were essentially two things that bothered Maria, three if you counted the food: what appeared to her to be "an arrogant flaunting of wealth," and a lack of interest in and knowledge about Germany on the part of many of the people she met. Maria says, "Everybody smiled a good deal and took a great many pictures, but you could never sit down for a serious discussion about anything. They didn't care about my being an Ossi [East German], although I must admit that if someone had told me they were from Colorado, I wouldn't have jumped up and down and congratulated him or asked a lot of questions. Still, I was amazed at how little they knew about Germany. If they did ask questions about the GDR, they were really naive. It was as if they had lived in a prison their entire lives. It is true that I didn't speak enough English to explain my situation or to express my thoughts, but even so. . . . What I did was to prepare simple answers for their simple questions. That made it easier for me to get along—especially on the seven-day tour, when we gave nine concerts and stayed with a different family every night."

The experience she enjoyed the most was staying with a couple in Birmingham. The husband was German. Maria says, "He was the first one to understand my culture shock. We spent time mimicking people, saying, 'You all have fun, now,' or 'Have a nice day.' The husband was almost frighteningly ugly, while the wife was good-looking, so I called them 'Beauty and the Beast.' They lived in a small apartment and were not rich, as were all the other people I stayed with, nor were they so religious. I really enjoyed being with them. They were like my mother's friends whom we stayed with on the other trip, who were university-affiliated and with whom I could have conversations. This time, I saw a different kind of people. They seemed to be primarily interested in making money and far less concerned with things going on outside their country."

Maria is of the opinion that her trip to the American South gave her a more balanced understanding of the United States. She says, "My mother had told me about the dark side of American life, and I felt this time that I saw what she meant."

Maria believes that she will always retain some of her earlier, magical feelings about America, but she is less certain that she will return in the near future to the United States as an exchange student.

GDR

If America has been overvalued in Maria's eyes, the GDR has been undervalued. It is not that she would want the GDR to return, but there were beliefs held then—such as the commandments she had to memorize as a Young Pioneer: belief in the community, in solidarity, in cooperation— that she thinks are needed now more than ever. She says, "In the GDR, all the kids played together. Now I can see how a new differentiation is taking place, one that excludes the poor ones. Back then, there was no Nintendo or other expensive games that could mark those kinds of differences. Today, everybody my age is trained in how to sell themselves, how to work up their résumés, how to prepare for job interviews. They are also being told not to object to certain ugly things that are happening in Germany and elsewhere if they want to get a job. Of course, in some ways it is not altogether different from the GDR—just a lot more so."

Maria says, "I have started talking to my parents more about those times, even though I do not believe everything that they have to say. I remember that up until I was twelve or thirteen, I believed my mother to be the guardian of absolute truth. I remember getting into fights with my friends about the first elections in 1990 and quoting my parents about what they believed. It was only later that I started questioning my parents, but even now I am mostly in agreement with them. My mother is more serious than my father. I guess that I'm somewhere in between."

Maria says that she does not really know how her parents felt about *die Wende*. She remembers them saying that they were optimistic in the beginning but later realized how an important historical opportunity had been missed. After that, it was clear that everything was to be modeled after the West. By October 3, 1990 (official reunification), they were disillusioned. As for her own response, Maria says, "I remember that I felt a certain defensiveness when people from the outside ridiculed the country at that time. Then, too, we were supposed to feel that everything was immediately better after the Wall came down. The question that was asked so often after *die Wende* was, 'So how do you like your new freedom?' But I never quite understood what the 'new freedom' was that they were talking about.

"I realize that I was exposed to a lot of propaganda when I was a child, but I still enjoyed living in that country, and I wanted to defend it—even at the end."

The West

When the Wall came down in 1989, Maria was eleven. She remembers going with her mother and father to Berlin to see the Wall while most of it was still there. At that time, she believed that West Germany was an entirely different country, like Poland or Hungary. It struck her as strange that they wanted to take down the border, even if they did speak the same language.

In 1992, Maria made her first extended trip to the West, traveling with her father and mother to Mosbach, near Heidelberg, for the wedding of a family friend. As they drove from town to town, noting the restored half-timbered houses (*Fachwerkhäuser*), the geranium-laden window boxes, and the clean streets, Maria had the distinct impression of traveling through a "Legoland," where real people did not live. She felt a tinge of homesickness for Göttern, with its muddy streets and unpainted houses.

Maria points out that among students, there is ongoing discussion about their teachers from the West. For many, their presence in the school is gratuitous. Why is it, the students want to know, that they import teachers from the West when there are so many good ones in the East? Disturbing to Maria is the fact that teachers from the West receive hardship bonuses (*Buschzulage*), because, as Maria says with exasperation, "They have to do service in the wild East!" On the other hand, Maria says, "There are quite a few 'neck twisters' [*Wendehälse*, or birds that can turn their heads 180 degrees] who, after *die Wende*, had no problem going opportunistically in the opposite direction after having previously been faithful socialist supporters. Personally, I prefer those East Germans who took up teaching after *die Wende*."

Coming of Age

Looking back on her childhood, Maria remembers having gotten her way a good part of the time. She never thought of her parents as being strict. Maria says, "My parents were softies. In their view, 'yelling was not instructional,' and spanking was out of the question. If something happened and I didn't come home when they expected me, they would tell me that they 'had been worried.' I don't know if they realized it or not, but that had more of an effect on me than if they had yelled, which my grandparents were more apt to do. Certainly the two generations are very different in that respect."

Maria admits to feeling shocked when her mother offered her, in a casual conversation when she was fourteen, the address of a gynecologist—just in case. Maria says, "My mother is pretty cool about those things. She basically trusts me. Even if I don't tell her everything about my love life, she just knows by looking at me that something has changed and that I won't be able to keep it all to myself. Whenever I stay over in Weimar with a male friend, just because I need a place to stay, I tell my parents, of course, but to Oma I always suggest that it is a girlfriend I am staying with."

For the present, Maria says that she is not thinking about getting married and starting a family. She says that she feels much too young for that: "It was just a little while ago that I was running around here like a kid."

Maria and the Schorcht Name

Maria has no overwhelming desire to leave Göttern, even though she knows that one day she will go away to the university. She likes the village; she likes their house. She only wishes that her grandfather would leave it alone and stop changing things around. If she had her way, she would have the plaster taken off the exterior of the house that he had put on some years ago to cover the old bricks. She particularly likes the garden that runs out behind the barn up toward the Gommel. She says, "I always want to stay a country girl somehow, because I know how fresh vegetables and fruit look and taste. Once you have drunk real cherry juice, you would never be satisfied with the kind you buy in a store. I would always want to keep certain trees and vegetables in the garden, though I would do without the heavier ones, like turnip and kohlrabi. All these garden and outdoor things are aspects of my family's history that

I would want to retain. There are others that I would not want to keep, like the way my forebears were very strict about persons from different social backgrounds getting married. They were pretty intolerant in these villages.

"I have thought quite a lot lately about the life my grandparents lived, how Opa had a kind of short childhood because the war started so soon. And how it didn't allow my grandmother to get more education. She wanted to have a profession; she has talked a good deal to me recently about that, how she missed out. Then the war stories . . . Opa must have had a lot of luck in the war. Most of his memories are about his wounds, but recently he has also talked about Hitler and the propaganda machine that Goebbels built. Then, too, how it was a good time, in the beginning, with the Hitler Youth and the games they played. Then after the war the Russians came—no, wait, it wasn't that way. The Americans came first—but it doesn't really matter. The important thing was that the Germans weren't in control anymore."

Sometimes, Maria says, her grandfather talks to her about the Jews and how they knew only the "good Jews" in Jena. He told Maria that because many of the Jews were intellectuals, the Germans probably envied them, which could be one of the reasons for the Holocaust (Maria uses this word, not Edgar). Maria says, "After the war, Wally came to Göttern, they married, then had three children. That was the 'reconstruction' time after the war. It sounds as if their lives were very monotonous, full of hard work. I do not envy their existence then. But now they are trying to get the most out of life that they can." After a pause, Maria continues, "I never actually talked to them about the GDR, but I didn't really think it was necessary. I already knew about the LPG [agricultural production cooperative]. For me, the LPG signifies the GDR in a certain way. I was born into the LPG—it was just there, and I never, ever wondered how it came about. You never wonder about those things that are just there—until they are not there anymore."

Maria says that she has little sentiment about being a German. She understands her father, she says, when he tells her that he does not feel like a German: "It was just by chance that I was born in Germany. All those nationalistic feelings just 'pass behind my back.' I feel the same as I did when we were in school and we had to sing a song about being proud to be Pioneers. We changed the lyrics from 'We are proud to be Pioneers' to 'We are proud to be sausages.' I dislike all that 'hysteria' business. It reminds me of what Opa and Oma have been talking to me about lately, how they enjoyed certain things in the early Hitler years, such as the time when Opa got all excited over Hitler speaking in front of the Elephant Hotel. I can't see how this is any more intelligent than a bunch of teenagers standing in front of a hotel, screaming their heads off for Michael Jackson."

Maria does feel strongly, though, about the Schorcht name. She is not certain, she says, whether she will keep it when she gets married; that depends on "his" name. She says, "It would be too bad to lose the Schorcht name, especially here in Göttern. If I keep it, though, it would make it more complicated for anthropologists to write a family history."

Commentary

Although the first twelve years of Maria's life were spent in the GDR, she now lives in the German Federal Republic. The change suggests that the young Maria physically moved from one nation-state to another, as thousands of Germans did in the nineteenth century when they migrated from their homeland to the United States. But Maria's migration was an internal one, as one Germany succeeded another in her childhood, just as Edgar and his father before him experienced living in four different Germanys while remaining in Göttern.

What Maria was "witness" to, beginning as an infant, was the progressive disengagement of the Schorchts from the LPG, and thus from the GDR, marked by her grandmother's retirement from the Magdala stables in 1981. Maria experienced further termination of the GDR when the Pioneers were disbanded, along with the FDJ, which she was about to join. Whenever the transition from one Germany to the other finally took place for Maria, the process of evaluation of one against the other began and still continues.

In her own mind and in "debates" with her parents, Maria keeps returning to what the GDR was, what they lost in its passing, and what they have in its place. Maria shares with her parents the status of being a child of a Germany that exists only in its name and in their memories.

It is clear in Maria's mind that she does not want the GDR back again. Even when she still lived in it, she expressed her impatience with its militaristic posturing and ideological fanaticism. However, its articles of faith having to do with cooperation and assistance—which Maria at one time took for granted—now assume, in their absence, a new significance. Many of her fellow students, Maria believes, take the new emphasis on individualism to heart at the cost of ignoring everything else. Sometimes Maria wonders whether her criticism of the post-GDR emphasis on self-promotion stems from her personal dislike of competition, but if so, she would only have to think of her grandfather Edgar, who, aggressive and competitive, shares with his granddaughter many of her concerns for a lost sense of community and cooperation.

The question that looms largest in Maria's mind, one that she wishes to discuss more at length with her parents, is why the future course of their country was modeled so completely on the West in 1990, forfeiting the redemptive aspects of their past.

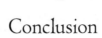

Conclusion

Part I

The New Göttern

MY ACCOUNT of the Schorcht family history draws to an end at a moment when much has come into flux—for the Schorcht household as well as for the village of Göttern itself. What had so long been taken for granted is now being called into question. The winds from the West, at first held at bay, are beginning to make their power felt in Göttern, as in all of Thuringia. It is not that everything that looms on the horizon is perceived as a threat but that so much is simply different.

The most visible change is the new Göttern arising. A rupture has occurred in the seemingly impermeable boundaries of the community. At first a visitor to the village does not notice it, for it is hidden from view on the road from Magdala by the curve of the Gommel. It is only when one passes the *Anger* and turns to the right, instead of driving on through the village in the direction of Bucha, that one sees the new settlement: a semicircular development of fifty single-family houses, arising directly beyond the former outer limits of the village to the southwest. The houses' dazzling white exteriors, redolent of similar ones in the West, stand in strong contrast to the dun-colored walls of the traditional dwellings of Göttern. They are being bought, for the most part, by professional and middle managers from Jena, eager to exchange the country for the city. Eschewing the proposal of a townhouse complex to have been built on the outskirts of Jena, they have uniformly chosen to partake of village life by investing in a new house in Göttern, Bucha, or Magdala, where similar developments are being erected.

A developer from Jena, Jena Invest GmbH, bought the land and built the necessary infrastructure: roads, lights, and sidewalks, with piped water and sewage for each lot. The cost per plot runs from 250,000 to 500,000 deutsche marks (DM) (as of 1995), depending on its size and location; land now costs 110 DM per square meter.

It is estimated that when the new settlement in Göttern is completed, the population will double, moving from its present size of 250 to 500. The prospect delights the owner of the newly renovated Gasthaus Zur Linde, the main local pub. He has already built himself a new house on the settlement. Others are less sanguine at the thought of Göttern becoming a bedroom town for Jena. Among other things, they are wondering what new services will be required.

Edgar rides over almost every day on his bicycle, just to keep an eye on what is happening. Maria, for one, is surprised that he approves of the new Göttern as much as he does, but then he always was one for changing things around.

Göttern Depoliticized

To minimize redundant efforts in these growing communities, the Common Municipal Administration (*Verwaltungsgemeinde*) was formed, bringing several neighboring communities under one administrative roof. According to new legislation for Thuringia passed on October 16, 1994, at the prompting of the State Interior Ministry, only communities with a population of five thousand or more could be independently administrated. Outraged at the prospect of losing its independence, Göttern only reluctantly, and under duress, accepted being incorporated into the Common Municipal Administration of the town of Magdala, in conjunction with the villages of Ottstedt and Maina. It is the incorporation into, rather than the affiliation with, Magdala that disturbs most people. The traditional office of the mayor has been reduced to the role of an unsalaried honorary official (*Ehrenamtlicher Ortsbürgermeister*) who, in conjunction with a village council, has jurisdiction over matters relating to culture and sports. Substantive administrative power is deflected to Magdala, while Göttern is represented by elected councilors who sit as members of a municipal assembly. The new official name of the village is Göttern Stadt Magdala. Still more demeaning is the letterhead designation: *Stadt Magdala, Ortsteil Göttern* ("The Town of Magdala, Subsection Göttern").

Few citizens—formerly participants in community affairs in Göttern—now attend council meetings in Magdala. There is widespread feeling in the village that after waiting for so long to be able to vote openly and directly for their immediate representatives in their own community, the new highly touted democratic freedoms bestowed on them have been betrayed in the name of enhancing efficiency and saving money. There is talk in Göttern about people "withdrawing from the whole process" and "letting Magdala worry about it." This perceived imposition of "foreign structures from the West" has been identified as a regressive administrative model deriving from supposedly conservative West German sources: Rhineland-Pfalz or Baden Württemberg. Whimsically, it has been suggested that Göttern has returned halfway to the conditions of its medieval existence, when it was a feudal appendage to the knights of Magdala. The paradox of a community that is doubling in size in a matter of a few years at the same time that it is losing its authority seems to some to be symptomatic of the situation eastern Germany finds itself in following the unification. As Heidrun puts it, "People do not have the energy to go to Magdala to ask questions. The steam has been taken out of Göttern. People are tired and stressed from fighting to make ends meet, or they have become slaves to their business and don't have time for anything else."

Edgar and Wally Experience New Worries for Their Children

Like others in Göttern, the Schorchts feel uncertain about the future. Apart from health problems typical of their age but exacerbated by the war and farm work, Edgar and Wally worry about their children. Edgar has been likened to a hen fretting over her brood, confronting problems that did not exist before *die Wende*. Foremost is the concern about employment. Edgar says, "We are concerned about our children. Will they keep their jobs?" Attainment of a secure position has been a constant source of anxiety since 1990 and, in the case of Erhard, even before. He was the first member of the family to be victimized by unification. Erhard had already been made redundant by 1989, when the project he had been working on at the University of Jena came to an abrupt end. Since then, he has worked as a freelance photographer and filmmaker, an unreliable source of income at best.

For her part, Heidrun is thankful to have her present position teaching German in Lithuania, where she earns a good salary. Heidrun feels confident that she can continue working in the high school at Panevėžys on a year-to-year basis, but for her own sense of well-being, she needs a permanent position in Germany—one that is not so far from her family, as her father's health is precarious.

To add to Edgar and Wally's concerns is the situation facing Roswitha and Manfred, both currently employed on the service staff of the Sophienhaus Hospital in Weimar. A new, four-hundred-bed hospital is being built in Weimar West, the Weimarer Hufelandkliniken, at a cost of 200 million DM, to replace the Sophienhaus and the Poliklinik. Not all personnel from both hospitals can be carried over, and it is too early to tell who will become redundant. Thus, there is only anxious speculation.

Norbert's job as an ambulance attendant is apparently secure; rather less so is Renate's job. Her former position as the administrative assistant to the mayor of Hetschburg, a job she treasured, came to an end when the office of salaried mayor was eliminated by the administrative reform of 1994. She is now serving as a nurse's attendant in Bad Berka, having had no nursing training. Norbert and Renate are relieved to know that their house, which Edgar helped them buy and renovate, is securely theirs.

As for the homestead in Göttern, Edgar's concern is not that it will be neglected by those who survive him but that it will become an imposition on his children. He says, "I don't want it to become their burden. There is no way they can make money from it, but it will need upkeep. Thank God we lowered the barn when we did. No one has the time or energy to do all that work now."

The Disposition of Edgar and Wally's Property

For some time, Edgar and Wally had been thinking of how to best dispose of the ownership of the Schorcht property. The transmission would be the seventh since Johann Heinrich came to Göttern from Söllnitz more than two hundred years ago. This transference would be different from all the others. Edgar was going to do something his forebears had never done: make the property partible, giving a portion to his daughter. Such a deviation from tradition became an inevitability when Edgar insisted on building an apartment for Roswitha and her family on the south side of the courtyard in 1987.

Edgar felt deeply the responsibility facing him. As far as he was concerned, he was swimming in uncharted waters, freed as he was of the expectations that had guided his forebears. During the two years after the unification, he made a point of consulting informed sources and attending presentations given on the subject of inheritance. He was determined to undertake the transmission properly. Heavy on his mind was the goal to do right by Roswitha to avoid an inequity similar to that forced on Irmgard. Norbert, the younger son, disinherited by reason of tradition, was also a concern. Edgar had already essentially provided him the house in Hetschburg—was that enough? Should he be given some land as well? Edgar consulted with his son on that point. Norbert's opinion, as reported by another member of the family, was that he had already received his share in the Hetschburg house; he did not want to cheat his brother and sister out of their fair allotment. Edgar also wished to make certain that his and Wally's right to reside in the house and be cared for until their respective deaths was assured.

It so happened that Stephen Kampmeier and I had our first interview with Edgar and Wally on the day before they were going to transmit the property in August 1992. The subject of this disposition came up in the course of talking about the history of the farm; obviously, it was much on Edgar's mind. He says, "Tomorrow we are going to the lawyers—I want to write the land over to my son and daughter. We divided things up and it is hard, because the courtyard is so small. Right now, it is OK. As long as we are alive, everything is OK. But after we are gone? Often brothers and sisters split up over things like that, and it is usually only a matter of money. Right now that is not the case, because my wife and I made an effort to provide a home for all the children. We paid for that ourselves." Edgar had made certain that the notary included a clause whereby Norbert in effect says, "I have received enough from you; I and my heirs renounce my claim to anything further." It is not that Edgar does not trust his children; rather, given the misunderstandings that inheritance can create, he believed it was best to spell out the contingencies from the start while he and Wally were still alive.

On August 18, 1992, Edgar, Wally, Erhard, Roswitha, and Norbert drove to the office of Dr. Wilhelm Hügel, notary, 30 Brühl Street in Weimar, to undertake something similar but also different from what Hugo, Edwin, his three sisters, and their husbands had accomplished seventy years before in the Blankenhain courthouse. The document, eighteen pages in length (as compared to the five-page 1922 instrument), was read with care by Dr. Hügel to those assembled.[1]

In brief, the first section lists the property to be transferred, essentially that which Hugo had passed on to Edwin, and Edwin to his only son, Edgar. The second section stipulates that Edgar give to Erhard the house on the western side of the courtyard, while the renovated apartment on the eastern side gets transferred to Roswitha. The barn at the end of the courtyard and the garden beyond goes to Erhard. The barn across the way and the garden beside it is allotted to Roswitha. The land is divided so that the larger amount centered in the 16-hectare Waltersberge field and the adjoining woodlots is apportioned to Erhard, while the smaller seventeen lots are given to Roswitha, including the choice 5-hectare Gommel. Edgar explained that a more balanced division of the land, given the surveying and related costs, would have been prohibitive.

The third section, considerably longer than the other two, states that in return, Erhard guarantees, through the lifetime of his parents, living space in his house to them both—namely, on the first floor, the kitchen, the small living room, the "good" living room, toilet, and bath, and on the next floor, the use of the bedroom one door to the left of the stairs, containing two windows overlooking the *Hof* (courtyard). Edgar and Wally are permitted to keep domestic pets as well as hens and rabbits. They are allowed use of the *Hof* and the garden. All expenses pertaining to upkeep and repairs are Erhard's responsibility. Costs for heat, water, electricity, chimney sweeping, and trash disposal are Edgar's obligation.

The next paragraph deals at some length with the care and treatment that Erhard and Roswitha promise to provide their parents in case of illness and infirmity. If the time should come when neither is able to clean that part of the house reserved for them or to care for their apparel, shoes, and bedclothes; to go to the drugstore; or to undertake other necessary tasks, it will fall to their children to do so for them. In addition, Erhard and Roswitha will be expected to prepare snacks and meals at the "usual" hour—that is, at those times Edgar and Wally have come to prefer. Edgar was careful to stipulate that "usual" refers to *Kaffee und Kuchen* as well as breakfast, lunch, and "evening bread."

The following section stipulates that if, within the next thirty years, property other than the dwellings is sold, the proceeds will be divided into four parts, one of which will go to the parents, and one-quarter to each of the three children. If the sale occurs after the death of both parents, the children will receive one-third apiece. Accordingly, Norbert would receive either a quarter or a third of the proceeds. Edgar, however, reserved the right for him or for Wally to take back property if any of it were sold without his or her consent, or to recover Erhard's or Roswitha's share if either should die before them.

The document was duly signed by Edgar Schorcht, Wally Schorcht, Roswitha Netz, Erhard Schorcht, and Norbert Schorcht, and then it was notarized, with copies sent to the land registry, the finance office in regard to the land transfer tax, and to Gotha in regard to the gift tax.

What distinguishes this document from those that preceded it is Edgar and Wally's intention to transmit the property in such a way that it would provide interrelated but separable living quarters for themselves and two of their married children and families. No longer constrained to perpetuate the descent of the farm intact from father to son, Edgar and Wally were able to identify another purpose for the homestead by including rather than excluding other family members, thus bringing to a close an age-old source of familial hurt and anger.

There is little reason to believe that Edgar and Wally feel personally diminished by having renounced ownership of the property over which they had reigned as farmer and farmer's wife for many years. They give, rather, the impression of taking part in a natural cycle of change. Edgar says, as if to clarify the new conditions of their lives, "Well, now we are without anything." His tone suggests a bittersweet recognition of how the end of life resembles its beginning.

The Disposition of Edgar's and Wally's Mortal Remains

In the same spirit of forethought with which Edgar and Wally have approached the settlement of the family property, they have talked with their children about what they would like done at the time of their deaths. They have made it known that they want to be cremated in Weimar. Since Elly's burial in 1963, cremation has become the practice. Edgar and Wally have their hearts set, though, on the service being held in the church in Göttern rather than in the chapel in Weimar. When Edwin died, the Göttern church, suffering from years of neglect, was deemed unsafe. Edgar, a member of the church council, hopes that the lengthy repairs will be completed in time. As for their ashes, they wish them to be interred immediately, rather than waiting for an additional service of commitment. Even though an ample space exists adjoining the gravesite of Edwin and Elly, there is no precedent for reserving grave plots. Neither Edgar nor Wally expresses a sentimental desire to lie next to their respective parents. They are ready to accept, they say, whatever comes their way.

Part 2

Something Missing

When I left behind my own life in Weimar and that of the Schorchts in Göttern in November 1995 to return to Amherst, Massachusetts, I found myself increasingly thinking about a disparity I sensed between the pervasive level of discontent that infuses much of eastern German discourse and the actual conditions of the people's existence. That there are reasons for disaffection is beyond question. The level of unemployment in what was the German Democratic Republic (GDR) is one of the highest in Europe, and the threat of job loss hangs heavily over the heads of persons who were accustomed to being severed from work only by retirement, disability, or death. Women feel deeply the diminution of the comprehensive system of child support held to have been every woman's natural right. Couples in mid-career complain of being exhausted as they attempt to keep pace on a treadmill fueled by competition. The retired and the elderly express a sense of being disregarded, while the young lament that there is no public or political purpose with which they can identify. Although what is a source of discontent for some is not necessarily a source of discontent for others, a shared articulateness about the drawbacks of the situation in which eastern Germans find themselves is common to many.

To some interested observers, though, the reality of existence in the eastern portion of the unified Germany does not seem to be as deplorable as public and private discourse would have one believe. Certainly to the eye there are numerous signs of dereliction: the entangled remains of what were once factories, cities blighted by soulless blocks of prefabricated apartment houses, the corroded surfaces of buildings suffering from thirty to forty years of neglect. Yet there is not the formidable evidence of human misery characteristic of so much American urban space: bag people sifting through trash cans, homeless sleeping in doorways, derelicts seeking handouts. Everywhere in eastern Germany, deterioration is less widespread than restoration. Cityscapes dense with cranes testify to the rate of reconstruction; houses on every block encased in scaffolding signal their rehabilitation, while the approaches to cities are assuming the telltale characteristics of the strip: State-of-the-art gas stations, furniture outlets, and supermarkets abound. One would think that eastern Germany was experiencing an economic boom, and so it is in respect to the comparative level of investment vis-à-vis the sixties, seventies, and eighties. How long this ascending economy will continue is uncertain, attuned as much of it is to consumerism and to the service sector. The state of the economy to the contrary notwithstanding, a visitor has the impression that something valued in GDR life is now missing.

It may be that eastern Germans feel too much aggrieved by a fait accompli—unification that seems more like an occupation than a reconciliation—to allow them to appear ready to accept their problematic status in a new Germany. The implied criticism of Wessis by Ossis made apparent by the recent political history of Weimar and Göttern is a case in point.

The former mayor of Weimar, a Christian Democratic Union (CDU) stalwart imported from the city of Fulda in the West after *die Wende*, undertook a strong campaign to maintain his seat in the 1994 election. Self-confident and assured of financial ties with Western finance, he had already won several lucrative projects for Weimar and was instrumental in acquiring for the city the honor of being the *Kulturhauptstadt Europas* 1999. It was expected that he would win reelection. To the surprise of most, and to the dismay of the CDU-controlled Thuringian administration, the opposition candidate—a former Socialist Unity Party of Germany (SED) mayor of Weimar, who ran a low-profile campaign as an independent—won. It may be that this upset can be adequately explained by the fact that there was no third alternative (Social Democrats, Bundnis 90/Grüne) or that the loser was a Wessi (West German) and the victor an Ossi (East German), who—because of and despite of his political past—seemed to many to be both more trustworthy and more accessible. Nevertheless, I believe that the election results point to larger issues.

What these may be is suggested by the fact that the defeated incumbent was viewed as the personification of an entrepreneurial manager dedicated to marketing Weimar as a favorable target for investment—a herald of the new world lying at Weimar's doorstep, to be entered into if its citizens had the vision to take advantage of the opportunity. The victor was perceived as holding out another vision to the citizens of Weimar, one that accepted the past, when surety and common purpose rather than risk taking and personal advantage were the order of the day. If the local candidate's victory can be seen as a manifestation of, among other things, GDR nostalgia (*Ostalgie*), then that regression goes deeper than a romantic yearning for the particularities of *Konsum* (cooperative) products, music, and Trabis. Closer to home was the situation that developed in Göttern following Edgar's retirement from the mayorship in 1987. His replacement was a member of the Liberal Democratic Party of Germany (LDPD), and though an Ossi, he had connections with the West. According to Doris Weilandt, his vice mayor, he was amenable to money-making schemes promulgated by fortune seekers from the West—for example, the creation of a waste dump in Göttern and the division of the village into commercial zones supported by a costly infrastructure of streets and sewage facilities. His theme was that to make money, one has to spend it. For the people of Göttern, the mayor's ideas were too ambitious, too expensive, too Western, and his ways were too authoritarian. In the summer of 1993, he was impeached.

Although unification provided a solution to the GDR nemesis of a severely flawed economy, it confronted the citizens of East Germany with the daunting question of the overall meaning of their existence. If life is to be regarded primarily as an individualized rather than a socially mediated process, then what significance is to be given to the place of communitarianism in East German history?

Edgar and Wally convey the sense that the sixties and seventies were heady days in Göttern, when almost everyone in the village had the welfare of the agricultural production cooperative (LPG) Zur Linde in common. Individual good was perceived, at least by the Schorchts, as depending on what was beneficial for the collective. Although it is difficult to make comparisons between the time before and that after *die Wende*, it seems that the people of Göttern at present enjoy a markedly enhanced standard of living when compared to that which existed prior to unification. Yet still there is a sense of wishing for the shared experiences that once connected them.

What is true of the Schorchts in this regard is not so much specific to the history of East Germany as it is common to Eastern Europe and western Russia as well, with the exception that individualism has been historically even less developed beyond Germany's eastern borders. In the West, the emphasis has been quite different, to the extent that individualism has evolved into the dominant social doctrine—most especially in the United

States, where little credence is given to communal priorities. Is that not a reason, then, why an American reader has difficulty understanding why the Schorchts finally accepted, as readily as they did, admittance into and compliance with the aims of the Zur Linde collective, and why now, as nominal individualists, they look back on those days of "togetherness" with some longing? It is not so much that eastern Germans are ungrateful for their liberation from the political oppression they experienced at the hands of the SED or for the greater economic well-being that they now enjoy but that they miss those collectivized arrangements and encounters that gave support and focus to their lives.

The Schorchts seem ambivalent about individualism. Their references to the new "elbow" society, in which everyone attempts to force his way to the head of the line, underline their skepticism. For the first time, Germans of the East are confronting a capitalist culture that their compatriots in the West have had two generations to adjust to and make their own.[2] The reassuring guidelines, omnipresent in the GDR, although often suspect, are gone, replaced by a new discourse idealizing an independence that is perceived more as a token of disregard than of liberation. The operative assumption is that in time, former East Germans will adjust to and internalize the new individuating norms. Even if they do not, the older, more recalcitrant generation with the longer memories will die off. Meanwhile, hope is placed on the younger, seemingly more flexible generation.

Erhard and Heidrun have a different perspective on the relationship of the present post-unification society to the former GDR than do Edgar and Wally. Born in the early 1950s, they took collectivism as much for granted as the air they breathed, the water they drank. Its passing, however has not had the same emotional impact on them as it has had for Edgar and Wally, who made the GDR their own by a major assertion of will. The parental generation had to make it work, for there was no alternative, no turning back to individual ownership. Just as acquired nationality often makes the greater patriot, so those who entered the GDR after birth may retain a larger affinity for it now that it is gone. But what Edgar and Wally look back on with regret for its passing is not the political creed of those times, which they disdained, but its social ethos. Erhard and Heidrun, however, knew as they came of age that to be a normal human being was to be a communist. They also understood that history and social evolution were on their side, for the imperfect socialism represented by the GDR was a necessary step in man's progress toward perfection.

Unlike her father-in-law, Edgar, with whom she worked closely when he was mayor and she a Community Council member, Heidrun assumes a more dispassionate regard for the lost "togetherness" of the GDR. For her, such sentiment diminishes the integrity of socialist belief that she understands has to do primarily with relations of production rather than with the idea of communitarianism. Nevertheless, Heidrun does make a case for the GDR's having been more socially advanced than the Federal Republic. "Bonds between people were closer, stronger, back then," she says. "People felt responsible for each other, they took an interest in each other, and that was something they only noticed when it was gone, after the Wall came down."

Edgar, unlike his daughter-in-law, was never a believer in socialism or communism, despite the efforts of the SED to cajole him or of his colleagues to the left to persuade him of its inevitable triumph. Rather, he loyally followed his father's footsteps into the Liberal Party, which adhered to the retention of a German national party, private property, and a free market. The LDPD provided a banner for him, affording exemption from doctrinal restraints imposed by the SED on party mayors as long as he voiced adherence to national goals. At the same time, it afforded him the freedom to pursue his own communitarian vision for Göttern. Edgar saw no inevitable conflict between the individual right of ownership on one hand and a communalistic ethos on the other. Edgar was not an ideologue; he relied on Heidrun to help him sound like one when the occasion required. Edgar's and Wally's expressions of regret for the sense of "togetherness" that was lost with the passing of the GDR is a subtext of a more generalized discourse, one grounded in the realities of the post–Cold War present, in which a communitarian critique decries the assault of privatism on traditions of mutualism.

What privileges the Schorchts' testimony is the fact that they knew intimately the challenges and rewards inherent in independent and in collective farming. As private farmers, they resisted the imposition of collectivization on them. When they could hold out no longer, they dedicated themselves wholeheartedly to its success. In time, Edgar became a severe critic of state cooperative agriculture, as it moved toward complete industrialization of the countryside. Wally resigned from the LPG, invalided by work in the communal barns. Yet despite their misgivings, they retained a warm memory of the manifestations of support and care they associated with the LPG's activities.

What credibility, given their history of past adherence to former totalitarian regimes, can be attributed to Edgar's and Wally's expressions of disquiet for the present—a disquiet shared by many who see signs in every direction of an "underlying decline in felt obligations and ties to others"?[3] Are their criticisms timely counsels against an advancing social implosion, or are they capitulations to the attraction of totalitarian conformity?

Within the boundaries of the discourse that revolves around the relative claims of individualism on one hand and communalism on the other, there is no more articulate defender of the former than George Kateb, professor of politics at Princeton University. His book *The Inner Ocean: Individualism and Democratic Culture* is an elegant condemnation of communitarianism, which, in Kateb's opinion, is a facilitator of docility, "a condition in which people unreluctantly accept being used and do so because they have been trained to do so."[4] Docility is in turn a facilitator of fascism. Of the qualities that critics of liberal society believe that people in such societies need but do not find—togetherness, discipline, group identity, and mutuality, all but mutuality are, in Kateb's judgment, complicit with docility.

In Edgar's "communitarian" vision, as represented by the "Cleaning Saturdays" and other "Join In" programs he created as mayor, togetherness was most critical. We might expect the converse of togetherness to be lack of attention paid to others rather than simply a sense of loneliness. "They don't care anymore," says Edgar, making a sweeping generalization about the current lack of people dropping in on others—especially the older people in their homes, as he used to do as mayor. Paying heed to those for whom he felt responsible was an obligation Edgar took to heart. His sense of accountability in this regard is close to what Kateb identifies as mutuality— to care about others. Mutuality is one of the four needs people seek but do not find in a liberal, individualistically oriented society. It is the only one of the four sentiments that Kateb claims to be nonculpable in the creation of docility.[5] Even so, there are few if any grounds for reconciliation between Kateb and Schorcht; Kateb's distinction between acceptable mutuality and unacceptable togetherness, together with the stringency of Emersonian self-reliance, which informs his understanding of individuality, sets the stakes too high to allow for that. Edgar's and Wally's generalized commentary on the political and economic climate they now confront, the anxious prospect of an unconcerned future in light of a fostering remembrance of a concerned past, can be viewed only as surrender to docility in Kateb's terms. But if Edgar's and Wally's sentiments are to be regarded as evidence of docility, then are not all communal processes—those democratically organized as well—to be perceived as manifestations of docility? In his paper, "George Kateb—The Last Emersonian?" Cornel West asks a similar question of Kateb when he writes, "How can we democratize ourselves, our societies, without forms of political solidarity and community loyalty?"[6] But even if we accept the reasonableness of West's caveat, there still remains the possibility that the Schorchts' commentary on the present and the past masks a revanchist hope for the status quo ante. I think this is not the case. What the Schorchts miss is the spirit of egalitarianism and sharing that characterized much of daily life, especially in the sixties and seventies, and not the political system that encapsulated it. I am of the opinion that Edgar's and Wally's notion of "togetherness" is held and felt on secular grounds, residing in personal experience—in Edgar's case, in the example set by his father in insisting that the farmhands, prisoners of war (POWs), apprentices, and family members break bread in one another's company; in the "neighborliness" that Edgar remembers as prevailing in Göttern before the war; and in the brigade work that he and Wally often enjoyed in the early LPG days.[7] So it is with Maria, who shares her grandparents' disquiet for the current culture of aggressive self-promotion.

What is at issue here is not the relative worth of individualism and communitarianism as ways of being in the world. Such an evaluation must come from others, not from an anthropologist narrating the history of an eastern German family. Rather, the point being made is to confirm how the representation of daily life may reveal the experiential side of a formal discourse, the subjective reality of which is usually omitted.

Commentary

The Savos and the Schorchts

Before leaving for home in November 1995, I flew from Frankfurt to Rome for a long weekend in Italy. On that Sunday, I drove with old friends, Guido and Talia Fabiani, from Rome to Sermoneta, to visit with Maria and Giacomo Savo for the day. Even then, after forty-five years of knowing them, I was unprepared for the surprises that a visit to the Savo household held in store for me. The sight of the house, as we approached, caught my breath. It had been painted a coral pink. With its red tile roof, flowering bushes, and palm trees, it looked like a villa in Coral Gables. As I alighted from the car and embraced everyone in sight, I asked after Massimo, Giacomo and Maria's second son. I was told that he was exercising his horse. I walked down the road incredulous, not that Massimo had learned to ride but at the distance traversed by the family, which the presence of a horse in their midst signified. Massimo approached me, leading a well-groomed chestnut mare. He wore jodhpurs and riding boots. After we greeted each other, Massimo nuzzled his head against the horse's sleek neck while I took his picture. He spoke with pride about his work in the frozen-food plant; of his son, Alessio, doing well in school; and about the new BMW for which he had recently traded in his Lancia.

Later at lunch, Illaria and Anna, the two children of Eleuterio and Paola (Giacomo and Maria's first son and his wife), joined us at the table. Illaria, now a young woman of seventeen, is in her next-to-last year in the *liceo classico* (liberal arts high school) in Latina. No one else in the family had arrived at that educational point, nor had even contemplated doing so. Having been the first to cross the divide between junior high (*scuola media*) and high school (*liceo*), Illaria is now determined to take the next leap, into the university, leaving behind the traditional working-class aspirations of her family.

Since I first came to know them in 1951, the Savos have been involved in a remarkable odyssey that has taken them further than their wildest dreams. They began their voyage encumbered, as were so many Italians living to the south of Rome, by an extensive insufficiency. They lacked money, property, skills, education, and, above all else, connections with the rich and powerful. Their desire to push ahead was fueled by a wish to move out of the depths of poverty, where they had suffered deprivation and shame. They struggled step by step to better themselves. Their journey is marked by the milestones of modern Italian material culture—bicycle, Vespa, Fiat 500, Fiat 1200—by which they can gauge their own ascent.

Theirs is the story of millions of Italians whose familial odysseys transformed Italy, just as Italy itself, in the throes of change, transformed its citizens in the years following World War II. Any one of countless familial narratives would have traced similar itineraries, beginning in a "timeless" Italian rural culture and arriving in a cosmopolitan industrial one, all in the course of a single lifetime.

The story of the Schorcht family is equally compelling, although the historical territory through which they voyaged is importantly different, marked not by a vertiginous ascent through socioeconomic space but by a steady march through a succession of reincarnated Germanys. Rather than a story of social transformation, the Schorcht history is one of adaptation to the demands of differing ideologies.

The Schorchts have owned property for generations and have known relative prosperity. The reassertion of their independence following the unification of Germany did not have the same catalytic effect as did Giacomo's acquisition of 3 hectares of land in 1958. The flight of the Savos from their miserable two-room dwelling in the village to a new house in the plain was an act of liberation from destitution. Freed from the constraints of impoverishment, they have not looked back to those times except in wonderment as to how far they have come.

In contrast, the past has had a different meaning for the Schorchts: that of pride in their former accomplishments, when the name Schorcht elicited widespread recognition as a superior farming family. Schorcht memories evince redemptive images of the neighborliness (*Gemütlichkeit*) experienced during the Third Reich before the war and the togetherness enjoyed in the work of the LPG in the GDR. The past is kept within reach as a means of confronting a problematic present and future.

Memories of the past suggest premonitions of the future, though the outlook is rather different from that in Italy. The Savos have thrown in their lot with the future on the assumption that, starting with zero, they have nothing to lose. Change is perceived as having worked to their benefit. As Maria Savo says with a broad smile, "Who would have thought that peasants like us would ever be living as if we were well-to-do?"

The Schorchts, with long memories of their past success as farmers, in conjunction with the visionary projections of utopian futures imposed on them, are more guarded about what lies ahead. On more than one occasion, reference was made to an inscription found on the old staircase during the reconstruction of the stables: *"Führer, wir halten durch!"* ("Führer, we'll persist!"). The writing is decorated with a swastika and signed by three men who had worked there. It is dated 1945. One of the three carpenters is still alive today. The existence of this unwelcome prophecy casts its shadow on the future, deepening its ambiguity, for there is always the unstated thought that what happened between 1933 and 1945 could happen again.

It is, however, in the landscape of moral accountability that the most strongly delineated differences between the experiences of the Schorchts and those of the Savos arise, dissimilarities made more instructive by the fact that the Germans and the Italians were allied in the war against the Western democracies and the Soviet Union.

The skeletons in the Savos' closet, markedly at variance with those of the Schorchts, reflect ordeals of shame perpetrated against family honor: Rosa Patella, Maria's mother, was born out of wedlock, and Maria's older sister, Teresa Patella, was married, during the anarchy of wartime, to a man already living in wedlock, bringing unbearable humiliation to the Patella household. Demands for accountability to authorities for whom they had no respect were regarded by the Patellas as amoral, prodding them to disobedience. Francesco Patella, Maria's father, evaded service in Mussolini's African campaigns of conquest to avoid leaving his growing family, while his wife, Rosa, sold produce on the black market to feed them. Later, Francesco joined the Fascist Party in the hope that its powerful patronage would procure him much-needed employment and abandoned the party when it failed to do so. When the war descended on Sermoneta in 1944, Francesco avoided the roundup of able-bodied men on the part of the German occupying forces and went into hiding. Following the surrender of the Badoglio government to the Allies in 1943, the Patellas became victims of German aggression and terror, forcing them to exist as outlaws. Their subsequent recounting of these events, tinged with pride and free of guilt, served as witness to their destiny.

The Schorchts' predilection, on the other hand, to render unto Caesar his due led to moral equivocations that touched on our work from time to time: the apologies for Hitler, *der Führer* ("The war was madness, little Germany against the whole world—at fault for this were 'the little Hitlers,'" says Irmgard; "Hitler built the Autobahn, solving the problem of unemployment," says Edgar); membership in the National Socialist German Workers' Party (NSDAP [Edgar's and Irmgard's insistence that their father joined the party only reluctantly toward the end of the war, when, in fact, he became a member on April 1, 1942]); the consensus of ignorance about Buchenwald ("We didn't know anything about Buchenwald, did we, Edgar?" says Irmgard, turning in her brother's direction, an assertion made questionable by contact with Elly's brother, Walter Michel, a Nazi functionary in Weimar); Edgar's contention that Jews are not Germans ("You know," he says, "they [Jews] are very good at this [merchandising], better than we Germans"); and the disparity between Edgar's relatively benign account of the war on the Eastern Front and the depiction of the barbaric behavior of the Wehrmacht in the current literature.

A number of recent books argue the case that rather than being a professional army isolated from political control, the Wehrmacht fighting in the Soviet Union was essentially a criminal organization, ideologically radicalized toward genocidal ends.[8] The case is made in such a way as to leave little room for the individual exceptions that Chapter 7, "Edgar at War," suggests Edgar represented. The question arises whether Edgar, traumatized by the horror of the events that he was witness to, if not a participant in, has "forgotten" them. The German psychoanalyst Margarete Mitscherlich has stated, "The Eastern Front and Serbia were scenes of the worst military carnage. Diaries and letters of soldiers in action there reveal, from the beginning of the campaign against the Soviet Union at the latest, everybody knew exactly what was going on," and "For many Germans the only way of coping with the total devaluation of ideas of national grandeur was either to forget the past altogether or to entertain as cloudy a vision of it as possible."[9]

The evidence suggests that Edgar is dissembling in regard to the situation that prevailed on the Eastern Front. Yet to propose that he is denying what he was witness to is unproductive, because it is undemonstrable. More to the point would be to entertain Edgar's "particularistic" war account in conjunction with other familial affirmations and evasions during the Hitler years as representative of the selective memory of a German family for whom justification of the status quo was paramount. Edwin's belief in the essential rectitude of the state and the military, one substantially shared by his children, involved the family in situations of moral ambiguity resulting from the degree to which the nation and its armies participated in criminal activities.

Paradoxically, it was the disinclination of the Savos/Patellas to obey their nation's laws at critical times that absolved them from complicity in immoral actions taken in the name of Il Duce, the king, and the party.

Final Things

When I paid my last visit to Göttern on October 31, 1995, to bid farewell to the Schorchts, I was struck by my reluctance to have my time there end. There would still be much to do on my return toward the completion of the writing, but this was to be the termination of my sojourn in Germany. It had not always been an easy time; there were long stretches of solitude in Weimar, and the weather was, for the most part, abominable. I was determined not to spend another winter there. Yet I knew that I would miss not only contact with the Schorcht family of Göttern but also my life in Weimar: the implacable omnipresence of history at every turn and the wealth and quality of its cultural offerings.

Notes

INTRODUCTION

1. Hayden White, "The Value of Narrativity in the Representation of Reality," *Critical Inquiry* 7, no. 1 (Autumn 1980): 9.

2. Clifford Geertz, *Works and Lives: The Anthropologist as Author* (Stanford, CA: Stanford University Press, 1988), 144–145.

3. C. Wright Mills, *The Sociological Imagination* (New York: Oxford University Press, 1959), 6.

4. Jacques Le Goff, *History and Memory*, trans. Steven Rendall and Elizabeth Claman (New York: Columbia University Press, 1992), xi–xii.

5. Yosef Hayim Yerushalmi, *Zakhor: Jewish History and Jewish Memory* (Seattle: University of Washington Press, 1982), 94.

6. Hans Vaget, "'Eliminationist Antisemitism': Daniel J. Goldhagen Constructs German Culture," 10.

7. Saul Friedländer, *Memory, History, and the Extermination of the Jews of Europe* (Bloomington: Indiana University Press, 1993), 126.

8. Despite the redemptive presence of the spirits of Goethe, Schiller, Herder, Liszt, and Gropius, reminders of the evil that was Nazism are clearly evident in Weimar. Hardly a month goes by that the renowned *Weimar Kultur Journal* does not publish an essay on Buchenwald and its relationship to the history of Weimar. Bernd Kauffmann's piece "Weimar und Buchenwald," published in April 1995 and commemorating the fiftieth anniversary of the freeing of the camp, is a prominent example.

9. Geoff Eley, "Labor History, Social History, *Alltagsgeschichte*: Experience, Culture, and the Politics of the Everyday—A New Direction for German Social History?" *Journal of Modern History* 61, no. 2 (1989): 317.

10. For a discussion of *Alltagsgeschichte* and the problem of "normalization," see Ian Kershaw, "'Normality' and Genocide: The Problem of 'Historicization,'" in *Reevaluating the Third Reich*, ed. Thomas Childers and Jane Caplan (New York: Holmes and Meier, 1993), 20–41.

11. Ibid., 23.

12. Henry Krisch, *The German Democratic Republic: The Search for Identity* (Boulder, CO: Westview Press, 1985), 5.

13. Mary Fulbrook, *Anatomy of a Dictatorship: Inside the GDR, 1949–1989* (New York: Oxford University Press, 1995), 19.

14. Stephen G. Fritz, *Frontsoldaten: The German Soldier in World War II* (Lexington: University Press of Kentucky, 1995), 210.

CHAPTER 1

1. Not only were there insufficient funds to allow Edwin to pay his sisters; the post-war German economy meant that he would have to wait twenty years to fully reimburse them.

2. See *Bürgerliches Gesetzbuch* (BGB) Fünftes Buch. Erbrecht Erster Abschnitt. Erbfolge 1924, Abschnitt 1 and 4.

3. Wolfgang Huschke, "Politische Geschichte von 1572 bis 1775," in *Geschichte Thüringens*, Fünfter Band, 1. Teil, 2. Teilband, *Politische Geschichte in der Neuzeit* (Cologne, Germany, 1982), 1.

4. See Hans Medick and David Warren Sabean, eds., "Part I. Family and the Economy of Emotion," in *Interest and Emotion: Essays on the Study of Family and Kinship* (Cambridge: Cambridge University Press, 1984), for a discussion of how property interests may structure emotions.

5. See Dieter Dünninger, *Wegsperre und Lösung. Formen und Motive eines dörflichen Hochzeitbrauches. Ein Beitrag zur rechtlich-volkskundlichen Brauchtumsforschung* (Berlin: De Gruyter, 1967), 281ff., for an alternative explanation of this custom.

6. Black was the appropriate color for wedding gowns for Protestant brides of Elly Schorcht's social position.

CHAPTER 2

1. *Gommel*, usually *Kummel* in Thuringia, refers to what is thought to be a Slavic gravesite. Kurt Bürger, *Göttern, eine anthropologische Untersuchung aus Thüringen* (Jena: Fischer, 1939), 25n27.

2. Jerome Blum, *The End of the Old Order in Rural Europe* (Princeton, NJ: Princeton University Press, 1978), 121.

3. Alan Mayhew, *Rural Settlement and Farming in Germany* (London: Batsford, 1973), 52.

4. In discussing the three-field system as a variation of open-field agriculture, I have relied extensively on Carl Dahlman's *The Open Field System and Beyond: A Property Rights Analysis of an Economic Institution* (Cambridge: Cambridge University Press, 1980). A compelling characteristic of Dahlman's treatment of the open-field system is his insistence on treating it as a coherent system adaptive to varying conditions rather than as an aggregation of discrete historical elements whose existence for more than a thousand years defies explication.

5. On the subject of the role of communality in the three-field system as it existed in the Germanic and Slavic regions of Europe, the German historian Werner Rösener, author of *Peasants in the Middle Ages* (Urbana: University of Illinois Press, 1992), writes, "The introduction of the common-field system forced the peasants of a village

to co-operate in arranging the collective use of the fallow as pasture, which meant that herdsmen had to be appointed and further steps had to be taken to guarantee the smooth working of the village economy. Especially since the eleventh century the new methods of crop cultivation contributed to the formation of the village as an economic co-operative. The repercussions on peasant mentality of the altered modes of life and work engendered a communal spirit among the village inhabitants and put an end to the privileged use of farmland. For although the individual parcels remained the property of each peasant, individual farming was increasingly subject to communal regulations" (51).

6. For a more extensive discussion of inheritance patterns in general, see Jack Goody, Joan Thirsk, and E. P. Thompson, eds., *Family and Inheritance: Rural Society in Western Europe, 1200–1800* (Cambridge: Cambridge University Press, 1976).

7. Leo Drescher, *Der Grund und Boden in der gegenwärtigen Agrarverfassung Thüringens* (Jena: Fischer, 1929), 44ff (see also map at end of book). See also Hans Patzke and Walter Schlesinger, eds., *Geschichte Thüringens, Erster Band: Grundlagen und frühes Mittelalter* (Cologne, Germany/Graz, Austria: Bohlau, 1968), 250ff.

8. See "Geburts-und Sterberegister der Gemeinde Göttern" at the church archive in Magdala, and the inheritance contract between Anna Magdalena Schorcht and her son Johann Heinrich Schorcht Jr. (1816) in file A40, Göttern, "Kataster/Grundstücksveränderungen 1816–1918," Kreisarchiv Weimar. See also document entitled "Recess vom 15. Juli 1794," which explains the absolution from tax duty of a piece of land jointly owned by the Stöckel sisters, in file A30, Göttern, "Frohne-/Erbzins-/Triftablösungen 1662–1870," Kreisarchiv Weimar.

9. Bürger, *Göttern, eine anthropologische Untersuchung aus Thüringen*, 16n12.

10. See the "Erb-und Alimentations Contract" (March 5, 1816) between Anna Magdalena Schorcht and her son Johann Heinrich Schorcht Jr. in file A40, Göttern, "Kataster/Grundstücksveränderungen 1816–1918," Kreisarchiv Weimar.

11. File A40, Göttern, "Kataster/Grundstücks-veränderungen 1816–1918," Kreisarchiv Weimar.

12. Professor Hartmut Wenzel, personal communication, from Hochschule für Architektur und Bauwesen, now Bauhaus-Universität Weimar.

13. See file A31, Göttern, "Ablösungsangelegenheiten mit Erbzins 1846–1860 "Ablösungsvertrag 1916," Kreisarchiv Weimar. The first document dates from May 1846.

14. Lienhard Rösler, "Landwirtschaft Thüringens im Mittelalter," in *Thüringer Blätter zur Landeskunde*, ed. Landeszentrale für Politische Bildung (Thüringen: Erfurt, 1994), 3.

15. See documents regarding the abolition of feudal dues in file A31, Göttern, "Ablösungsangelegenheiten mit Erbzins 1846–1870—Ablösungsvertrag 1916," Kreisarchiv Weimar.

16. Ulrich Hess, "Geschichte in der Epoche des Feudalismus und des Kapitalismus," in *Der Landkreis Weimar. Eine Heimatkunde. Erster Teil: Allgemeines*, ed. Manfred Salzmann (Weimar, Germany: Stadtmuseum Weimar, 1980), 102f.

17. See "Ablösungsquittung, Altenburg 24.9. 1855" and "Ablösungsrevers Altenburg 17.10. 1856" in file A30, Göttern, "Frohne-/Erbzins-/Triftablösungen 1662–1870," Kreisarchiv Weimar.

18. Hess, "Geschichte in der Epoche des Feudalismus und des Kapitalismus," 104.

19. Heinz Haushofer, *Ideengeschichte der Agrarwirtschaft und Agrarpolitik im deutschen Sprachgebiet* (Munich: Bayrischer Landwirtschaftsverlag, 1958), 1:387. See also Hess, "Geschichte in der Epoche des Feudalismus und des Kapitalismus," 105.

20. James J. Sheehan, *German History, 1770–1866* (Oxford: Oxford University Press, 1989), 480.

21. Haushofer, *Ideengeschichte der Agrarwirtschaft und Agrarpolitik im deutschen Sprachgebiet*, 1:387. See also Blum, *The End of the Old Order in Rural Europe*, 119; Hess, "Geschichte in der Epoche des Feudalismus und des Kapitalismus," 105; and Sheehan, *German History*, 475–476.

22. See file A32, Göttern, "Separationsakten, 1862–1873," Kreisarchiv Weimar. See also file A34, Göttern, "Recess von Göttern und Gauga/Tabellarische Aufstellung der speciellen Plananweisung," Kreisarchiv Weimar.

23. See "Grund-und Alimentationsvertrag," dated August 20, 1847, which lists the Schorcht property and the inheritance contract, "Erb-Abtretungs-und Alimentationsvertrag," Göttern, April 15, 1893 (both documents in possession of the Schorcht family).

24. See file A32, Göttern, "Separations Akten 1862–73," Kreisarchiv Weimar.

25. Blum, *The End of the Old Order in Rural Europe*, 263.

26. Ibid., 257.

27. Sheehan, *German History*.

28. Ibid., 752–753. For further discussion regarding the introduction of chemical fertilizer in German agriculture in the nineteenth century, see Werner Rösener, *Die Bauern in der europäischen Geschichte* (Munich: C. H. Beck, 1993), 244.

29. Sheehan, *German History*, 753.

30. Walter Arnold and Fritz H. Lamparter, *Friedrich Wilhelm Raiffeisen, Einer für Alle—Alle für Einen* (Neuhausen-Stuttgart: Hanssler-Verlag 1985), 65–67.

31. Theodor von der Goltz, quoted in Sheehan, *German History*, 753. For further information in regard to von Goltz's perception of German agriculture in this period, see Theodor Freiherr von der Goltz, *Geschichte der deutschen Landwirtschaft* (Stuttgart: J. G. Cotta, 1903).

32. Sarah Rebecca Tirrell, *German Agrarian Politics after Bismarck's Fall: The Formation of the Farmers' League* (New York: Columbia University Press, 1951), 20.

33. Gordon A. Craig, *Germany: 1866–1945* (New York: Oxford University Press, 1978), 277.

34. See "Erb-, Abtretungs- und Alimentations Vertrag," Göttern, April 15, 1893 (in possession of the Schorcht family).

35. Rösener, *Die Bauern in der europäischen Geschichte*, 245. See also Bürger, *Göttern, eine anthropologische Untersuchung aus Thüringen*, 30, in regard to the decline of sheep raising.

36. Robert G. Moeller, ed., *Peasants and Lords in Modern Germany: Recent Studies in Agricultural History* (Boston: Allen and Unwin, 1986), 148–149.

37. Gustavo Corni, *Hitler and the Peasants: Agrarian Policy of the Third Reich, 1930–1939* (New York: St. Martin's Press, 1990), 1–4.

38. Robert G. Moeller, *German Peasants and Agrarian Politics, 1914–1924: The Rhineland and Westphalia* (Chapel Hill: University of North Carolina Press, 1986), 68–69.

39. Gitta Günther and Lothar Wallraf, *Geschichte der Stadt Weimar* (Weimar, Germany: Böhlau, 1976), 520.

40. Gerald Feldman, *The Great Disorder: Politics, Economics, and Society in the German Inflation, 1914–1924* (New York: Oxford University Press, 1993), 188–193.

41. Friedrich Lütge, *Die mitteldeutsche Grundherrschaft und ihre Auflösung* (Stuttgart: G. Fischer, 1957), 266–282.

42. "Kötschau vor der Separation. Zeichnung H. Wenzel nach der Flurkarte von 1863," Hochschule für Architektur und Bauwesen, Weimar.

CHAPTER 3

1. Hans Medick and David Warren Sabean, eds., *Interest and Emotion: Essays on the Study of Family and Kinship* (Cambridge: Cambridge University Press, 1984), 9.

CHAPTER 4

1. Karl Kautsky, *The Agrarian Question: In Two Volumes* (London: Zwan, 1988), 1:20.

2. Friedrich Behn, *Die Entstehung des deutschen Bauernhauses* (Berlin: Akademie-Verlag, 1957), 36.

3. Oskar Schmolitzky, *Das Bauernhaus in Thüringen* (Berlin: Akademie-Verlag, 1968), 12–14.

4. Ibid., 26.

5. Ibid., 15–18.

6. For additional information on *Fachwerk*, see Günther Binding, Udo Mainzer, and Anita Wiedemann, *Kleine Kunstgeschichte des deutschen Fachwerkbaus* (Darmstadt, Germany: Wissenschaftliche Buchgesellschaft, 1975).

7. Every religious institution in Germany is allowed by the Constitution to levy taxes on its members. See Article 140 in *Grundgesetz für die Bundesrepublik Deutschland* from May 23, 1949.

8. Gustavo Corni, *Hitler and the Peasants: Agrarian Policy of the Third Reich, 1930–1939* (New York: St. Martin's Press, 1990), 104 and 114n87; see also 260–264.

CHAPTER 5

1. *Max und Moritz* (1865) is a well-known German story for children by Wilhelm Busch (1832–1908). The "two prankster kids" (*die beiden bösen Buben*), as they are known in the German tradition, are the direct forerunners of *The Katzenjammer Kids*.

2. Fritz Sauckel was born on October 27, 1894. Trained as an engineer, he joined the Nazi Party in 1921. In 1927, he was appointed NSDAP Gauleiter of Thuringia, and in 1933, its governor. Sauckel was influential in establishing the Buchenwald concentration camp in 1937 and in making Weimar a Nazi Party center. In 1942, he was made plenipotentiary general for labor mobilization. He was convicted in Nuremburg for war crimes and hanged on October 16, 1946.

3. See Michael Burleigh and Wolfgang Wippermann, *The Racial State: Germany 1933–1945* (Cambridge: Cambridge University Press, 1991), 52.

4. There are many references to Hitler's racism; see, among others, Adolf Hitler, *Mein Kampf* (München: Franz Eher Nachf., 1925), especially chap. 11, "Nation and Race"; Burleigh and Wipperman, *The Racial State*, 37–43; George L. Mosse, *The Crisis of German Ideology: Intellectual Origins of the Third Reich* (London: Weidenfeld and Nicolson, 1966), 294–311; and Gustavo Corni, *Hitler and the Peasants: Agrarian Policy of the Third Reich, 1930–1939* (New York: St. Martin's Press, 1990), 19, 34n6.

5. For further information about Darré and the "blood and soil" policy, see Chapter 6.

6. Bürger, *Göttern, eine anthropologische Untersuchung aus Thüringen*, foreword.

7. Venus's remark about the need for children to remain in the countryside is in keeping with Darré's remark that "farmers' daughters should not 'abandon the flag' and not 'leave their mothers in desperate straights' but resolutely struggle on, especially in these difficult times, and stay in the countryside to fulfill patriotic duty"; Corni, *Hitler and the Peasants*, 234.

8. The Hotel Elephant, a hostelry since 1561, has been frequented by personalities from Goethe to Thomas Mann. Mann renewed the hotel's fame via his novel *Lotte in Weimar*. Hitler, a frequent visitor, had the hotel remodeled to its present state in 1938.

9. Gitta Günther, Wolfram Huschke, and Walter Steiner, *Weimar: Lexikon zur Stadtgeschichte* (Weimar, Germany: H. Bohlau, 1993), 321–322.

10. Ibid., 322.

11. Ibid.

12. Bürger, *Göttern, eine anthropologische Untersuchung aus Thüringen*, pref.

13. Ibid., 72.

14. Ibid., 78.

15. Ibid., 78–79.

16. Corni, *Hitler and the Peasants*, 27–28.

CHAPTER 6

1. Jürgen W. Falter, *Hitlers Wähler* (Munich: C. H. Beck Verlag, 1991), 256. See also J. E. Farquharson, "The Agrarian Policy of National Socialist Germany," in Robert G. Moeller, ed., *Peasants and Lords in Modern Germany: Recent Studies in Agricultural History* (Boston: Allen and Unwin, 1986), 235–236; and Gustavo Corni, *Hitler and the Peasants: Agrarian Policy of the Third Reich, 1930–1939* (New York: St. Martin's Press, 1990), 14.

2. Falter, *Hitlers Wähler*, 256–266. See also Corni, *Hitler and the Peasants*, 24–25.

3. See file A16, Göttern, "Gemeinde Göttern, Reichstagswahl am 14. September 1930," Kreisarchiv Weimar.

4. Corni, *Hitler and the Peasants*, 23.

5. Ibid., 22.

6. Karl Dietrich Bracher, *The German Dictatorship: The Origins, Structure and Effects of National Socialism*, trans. Jean Steinberg (Middlesex, UK: Penguin University Books, 1973), 198–199.

7. Ibid., 198.

8. Corni, *Hitler and the Peasants*, 25–26.

9. Falter, *Hitlers Wähler*, table 25.

10. See file A16, Göttern, "Gemeinde Göttern, Reichstagswahl 31. Juli 1932"; Gemeinde Göttern, Reichspräsidentenwahl, II. Wahlgang 10. April 1932"; and "Gemeinde Göttern, Reichstagswahl 6. November 1932," Kreisarchiv Weimar.

11. Gordon A. Craig, *Germany: 1866–1945* (New York: Oxford University Press, 1978), 558. See also Falter, *Hitlers Wähler*, table 31.

12. File A16, Göttern, "Gemeinde Göttern, Reichspräsidentenwahl, II. Wahlgang, 10. April 1932," Kreisarchiv Weimar. See also Craig, *Germany: 1866–1945*, 558.

13. Falter, *Hitlers Wähler*, 38 and table 25.

14. File A16, Göttern, "Gemeinde Göttern, Reichstagswahl 5. März 1933." See also Falter, *Hitlers Wähler*, 40.

15. File A16, Göttern, "Meldung der Gemeinde Göttern für die Reichstagswahl und Volksabstimmung am 12. November 1933." See also Helmut M. Müller et al., *Schlaglichter der deutschen Geschichte* (Mannheim: Meyers Lexiconverlag, 1990), 265–266.

16. Jeremy Noakes and Geoffrey Pridham, eds., *Nazism 1919–1945, Vol. 2: State, Economy and Society 1933–1939* (Exeter, UK: University of Exeter Press, 1984), 318.

17. Ibid.

18. Ibid.

19. John Bradshaw Holt, *German Agricultural Policy, 1918–1934: The Development of a National Philosophy toward Agriculture in Postwar Germany* (Chapel Hill: University of North Carolina Press, 1936), 206.

20. Ibid.

21. Ibid.

22. Ibid.

23. Farquharson, "The Agrarian Policy of National Socialist Germany," 239. See also Corni, *Hitler and the Peasants*, 153n6.

24. Corni, *Hitler and the Peasants*, 250.

25. Farquharson, "The Agrarian Policy of National Socialist Germany," 239.

26. Noakes and Pridham, *Nazism 1919–1945*, 323.

27. File A10, Göttern, "Gemeinde Göttern, Niederschriften der Gemeinderatssitzungen 1932 bis 1945," Kreisarchiv Weimar.

28. Staatsarchiv Weimar, Bürgermeistergeschäfte Göttern 1944, letters to the NSDAP-Kreisleitung from February 16, 1944, February 23, 1944, and March 7, 1944.

29. Ibid., oath of office of the mayor of Göttern on April 18, 1944.

30. Edward L. Homze, *Foreign Labor in Nazi Germany* (Princeton, NJ: Princeton University Press, 1967), 48. For further information regarding the use of Polish POWs as farm laborers in the Third Reich, see 23–44, 164–165, 287–287, and 301–302.

31. For more information about village spinning bees, see Hans Medick and David Warren Sabean, eds., *Interest and Emotion: Essays on the Study of Family and Kinship* (Cambridge: Cambridge University Press, 1984), chap. 14, and Farquharson, "The Agrarian Policy of National Socialist Germany," 250.

32. Farquharson, "The Agrarian Policy of National Socialist Germany," 250.

33. Konrad H. Jarausch and Gerhard Arminger, "The German Teaching Profession and Nazi Party Membership: A Demographic Logit Model," *Journal of Interdisciplinary History* 20, no. 2 (1989): 225.

34. Ibid., 208.

35. William Sheridan Allen's description of the rapid acceptance of Nazism in the school system of the town of Northeim (Niedersachsen) following Hitler's coming to power affirms the contention that Nazism and education were perceived as readily compatible. He writes, "More than any other institution in Northeim, the schools became active instruments of Nazism." William Sheridan Allen, *The Nazi Seizure of Power: The Experience of a Single German Town, 1922–1945* (New York: Franklin Watts, 1984), 259.

CHAPTER 7

1. The German Army not only exploited all available Soviet economic resources but also forced civilians to work at the front as well as in the Reich. See Omer Bartov, *The Eastern Front, 1941–45: German Troops and the Barbarisation of Warfare* (New York: St. Martin's Press, 1986), 137–138.

2. For further information on the demoralization of German troops on the Eastern Front, see Marlis G. Steinert, *Hitler's War and the Germans: Public Mood and Attitude during the Second World War* (Athens: Ohio University Press, 1977), 299–300. See also Bartov, *The Eastern Front*, 19–30, 35, 37, 91, 144.

3. The *Wilhelm Gustloff* was a *Kraft durch Freude* (strength through joy) ship, built in Hamburg in 1938. See William H. Miller, *German Ocean Liners of the 20th Century* (Wellingborough, UK: Patrick Stephens, 1989), 168. The sinking of the *Wilhelm Gustloff* was the most costly disaster in terms of human life in marine history. The ship was named after Wilhelm Gustloff, a Nazi martyr. Gustloff, born in Germany, moved to Davos, Switzerland, in 1917 and joined the Nazi Party in 1929. In 1932, Gustloff was appointed head of the party's "Foreign Organization" in Switzerland. On February 3, 1936, he was shot to death by a Jewish student. Hitler himself eulogized Gustloff at his funeral.

4. In "The Distortion of Reality," in *Hitler's Army: Soldiers, Nazis, and War in the Third Reich* (New York: Oxford University Press, 1991), 73, Omer Bartov represents the fighting on the Eastern Front as one of spiraling barbarity, characterized by "brutal enforcement of discipline, the intensification of partisan activity [resulting in] increasingly barbarous and indiscriminate retaliation by the army. This accelerating process of radicalization, visible at all levels of the *Ostheer*, reflected the true essence of the army in the East, but also of the army as a whole." According to his personal account, Edgar opted out of that fatal circularity.

5. In both *Hitler's Army* and his earlier book, *The Eastern Front*, Bartov stresses the degree of increasing susceptibility of German soldiers on the Eastern Front to ideological indoctrination as their military situation became more desperate and their behavior ever more barbaric. See especially the chapter "Indoctrination and the Need for a Cause," in *The Eastern Front*. On the contrary, Edgar referred to a diminution of his support for the war as it drew to a close. For additional confirmation concerning the criminal behavior of German soldiers on the Eastern Front, see Hannes Heer and Klaus Naumann, eds., *Vernichtungskrieg: Verbrechen der Wehrmacht 1941 bis 1944* (Hamburg: Hamburger Edition, 1995), especially "*Die Behandlung der verwundeten sowjetischen Kriegsgefangenen,*" by Christian Steet, 78–90, and "*Auf dem Weg nach Stalingrad, Die 6. Armee 1941/42,*" by Bernd Boll/Hans Safrian, 260–296.

CHAPTER 8

1. Charles B. MacDonald, *The Last Offensive*, United States Army in World War II: European Theater of Operations Series (Washington, DC: Government Printing Office, 1973), 381.

2. Robert T. Murrell, *Operational History, 80th Infantry Division* (Oakmont, PA: Robert T. Murrell), 216.

3. See "Protocol of Proceedings of the Crimea, 'Yalta' Conferences, February 11, 1945" and "Protocol on Zones of Occupation in Germany and Administration of the 'Greater Berlin Area' approved by the European Advisory Commission, September 12," in U.S. Department of State, *Documents on Germany, 1944–1985* (Washington, DC: Government Printing Office, 1995).

4. Christine Schäfer, *Die Evakuierungstransporte des KZ Buchenwald und seiner Aussenkommandos* (Weimar, Germany: Nationalen Mahn- und Gedenkstätte Buchenwald, 1983), 8.

5. Edgar Schorcht's letter, as mayor, to FDJ and Pioneer members of the Edwin-Hoernle-Oberschule, Milda, March 19, 1985.

6. Daniel Jonah Goldhagen, *Hitler's Willing Executioners: Ordinary Germans and the Holocaust* (New York: Knopf, 1996).

7. Ibid.

CHAPTER 9

1. Heinz Haushofer, *Ideengeschichte der Agrarwirtschaft und Agrarpolitik im deutschen Sprachgebiet,* vol. 2, *Vom ersten Weltkrieg bis zur Gegenwart* (Munich: Bayrischer Landwirtschaftsverlag, 1958), 356–360. See also H. Rosenburg's note in Hanna Schissler, "The Junkers: Notes on the Social and Historical Significance of the Agrarian Elite in Prussia," in *Peasants and Lords in Modern Germany: Recent Studies in Agricultural History,* ed. Robert G. Moeller (Boston: Allen and Unwin, 1986), 36, 37.

2. Gerhard Rudolph, "Wirtschaft," in *Der Landkreis Weimar: Eine Heimatkunde, Erster Teil, Allgemeines,* ed. Manfred Salzmann (Weimar, Germany: Stadtmuseum Weimar, 1980), 35.

3. Ibid., 35–37.

4. Gregory W. Sandford, *From Hitler to Ulbricht: The Communist Reconstruction of East Germany, 1945–46* (Princeton, NJ: Princeton University Press, 1983), 83.

5. Ibid., 99.

6. List submitted to county commission by mayor of Göttern. See file 85, "Bodenreform"—Kreiskommission, Weimar Ortskommission Göttern, Staatsarchiv Weimar.

7. Ibid. Letter from local to county commission, September 29, 1945.

8. Ibid. Letter from Alfons Becker to county commission, September 28, 1945.

9. Ibid. Report on Wagner being taken into custody, October 10, 1945.

10. Ibid. Report on decision to divide state domain, October 17, 1945.

11. Ibid. Letter from Mayor Rothe and Walter Schulze to police, October 11, 1945.

12. File 1, Landesbodenkommission. Berichte der Kreiskommission, Weimar Staatsarchiv Weimar. Report of visit of Herr Ritter of county commission to Göttern, October 10, 1945.

13. File 85, Staatsarchiv Weimar. Letter from W. Schulze to county commission, October 15, 1945.

14. Ibid. Complaint from Kley, Schmidt, Sommer to county commission, October 18, 1945.

15. Ibid. Letter from Günther Müller to county commission, October 28, 1945.

16. Ibid. Certified copy of general assembly meeting, October 27, 1945.

17. Ibid. Registered mail from local commission to county commission, October 8, 1945.

18. Ibid. Registered letter from Becker to county commission, November 11, 1945.

19. Ibid. Certified copy of report of general assembly meeting, regarding division of state domain, October 27, 1945.

20. Ibid. Allocation of the livestock of the state domain, Göttern, December 8, 1945.

21. File 162, Landesbodenkommission. Declaration of party nonmembership, Staatsarchiv Weimar.

22. File 1, Landesbodenkommission. Protocol translated from Russian into German, USSR Captain Rowensky, November 23, 1945.

23. Rudolph, "Wirtschaft," 37.

24. Ibid., 36–37.

25. File 334–335, Landesbodenkommission. "Work Report of the VdgB, 1945," Staatsarchiv Weimar, including the first VdgB working guidelines.

26. Ibid. Establishing of machine lending stations.

27. Ibid. Distribution of equipment.

28. File 336–338, Landesbodenkommission. Memo from VdgB State Farmers Secretary to VdgB County Farmers Secretaries, "May 1st," October 18, 1947.

29. File 334–335, Landesbodenkommission. Comment on breeding station, Staatsarchiv Weimar.

30. File 55, Landesbodenkommission. Minutes of county commission, in which Becker supports expulsion.

31. SMAD decree 209, mentioned in Rudolph, "Wirtschaft," 37. See also file 334–335, Landesbodenkommission. Building program, November 15, 1947, Staatsarchiv Weimar.

32. Professor H. Wenzel, personal communication.

33. File 334–335, Landesbodenkommission. Letter from Thuringian Ministry of the Interior to the county commission, dated January 19, 1948. Staatsarchiv Weimar.

34. Göttern, "Municipal law of the town of Göttern concerning service of manual labor and hauling," November 7, 1947.

35. File 1, Kreiskommission. Report from county commission to Soviet authorities in Weimar, August 25, 1948, Staatsarchiv Weimar.

36. File 341–334, Kreiskommission. Report on furniture, Staatsarchiv Weimar.

37. File 334–335, Kreiskommission. Report of completion of houses for 1948; Thuringia goal exceeded by 9%, Staatsarchiv Weimar.

38. Sandford, *From Hitler to Ulbricht*, 113. See also Arnd Bauernkämper, "Von der Bodenreform zur Kollektivierung. Zum Wandel der ländlichen Gesellschaft in der Sowjetischen Besatzungzone Deutschlands und DDR 1945–1952," in *Sozialgeschichte der DDR*, ed. Hartmut Kaelble, Jürgen Kocka, and Hartmut Zwahr (Stuttgart: Klett-Cotta, 1994), 119–143.

39. Volker Klemm et al., *Agrargeschichte: von den bürgerlichen Agrarreformen zur sozialistischen Landwirtschaft in der DDR* (Berlin: Dt. Landwirtschaftsverl., 1978), 170–171.

40. Henry Krisch, *The German Democratic Republic: The Search for Identity* (Boulder, CO: Westview Press, 1985), 11–16. See also Müller et al., *Schlaglichter der deutschen Geschichte*, 346–347. For a more theoretical consideration, see Lutz Niethammer, "Erfahrungen und Strukturen: Prolegomena zu einer Geschichte der Gesellschaft der DDR," in Kaelble, Kocka, and Zwahr, *Sozialgeschichte der DDR*, 112–113.

41. Sandford, *From Hitler to Ulbricht*, 114. For a different view on the results of the land reform program, see Klemm et al., *Agrargeschichte*, 167–169.

42. File 363–364, Landesbodenkommission. Information on land returned, September 25, 1950, Staatsarchiv Weimar.

43. File 334–335, Landesbodenkommission. On larger machines designated for joint use, Staatsarchiv Weimar. For further information on machine lending stations, see Klemm et al., *Agrargeschichte*, 167–169.

CHAPTER 10

1. "Directive to the Commander-in-Chief of the United States Forces of Occupation Regarding the Military Government of Germany, May 10, 1945," in U.S. Department of State, *Documents on Germany, 1944–1985* (Washington, DC: Government Printing Office,

1995), section 6, "Denazification," 18–23. For a critique of the enforcement of the program, see Clemens Vollnhals, ed., *Entnazifizierung: Politische Säuberung und Rehabilitierung in den vier Besatzungszonen 1945–1949* (Munich: Deutscher Taschenbuch Verlag, 1991).

2. Tehalit GmbH in Heltesberg, Rhineland-Palatinate, now produces, among other components, electric-installation channel systems and elements for the European Installation EIB.

3. Irmgard was not trying so much to justify herself politically to her family as to everyone in Göttern, who, at least officially, regarded working for capitalists as culpable.

4. Gordon Craig makes reference to Minister of Education von Raumer's directive to all teachers to "impress upon their charges discipline, order and obedience to authority." Gordon A. Craig, *Germany: 1866–1945* (New York: Oxford University Press, 1978), 188–189.

5. Ingeborg Weber-Kellermann, *Saure Wochen, Frohe Feste: Fest und Alltag in der Sprache der Bräuche* (Munich: Bucher, 1985), 29, 144.

6. File A16, "Wählerliste zur Wahl des Gemeinde und Kreistages am 10. September 1922 in der Gemeinde Göttern." Kreisarchiv Weimar.

7. "Aufnahmeerklärung in die NS-Frauenschaft," Elly Marie Marta Schorcht, am 10. February 1934, Bundesarchiv, Aussenstelle Zehlendorf, Wasserkäfersteig 1, Berlin.

8. Gerhard Wilke, "Village Life in Nazi Germany," in *Life in the Third Reich*, ed. Richard Bessel (Oxford: Oxford University Press, 1987), 22.

CHAPTER 11

1. Klemm et al., *Agrargeschichte*, especially the following chapters: "Historische Bedeutung der 2. Parteikonferenz der SED" and "Der Leninsche Genossenschaftsplan," 173–181. According to Lenin in "Entwurf des Programms der KPR (B)," Werke Bd. 29 (Berlin: Dietz Verlag, 1961), 123: "In Anbetracht dessen, dass der Gegensatz zwischen Stadt und Land eine der tiefsten Ursachen der wirschaflichen und kulturellen Rückständigkeit des Dorfes ist . . . sieht die kommunistische Partei . . . in der Aufhebung dieses Gegensatzes eine der Grundaufgaben des kommunistischen Aufbaus."

2. Hanns Werner Schwarze, *The GDR Today: Life in the "Other" Germany*, trans. John M. Mitchell (London: Wolff, 1973), 54.

3. Gerhard Rudolph, "Wirtschaft," in *Der Landkreis Weimar: Eine Heimatkunde, Erster Teil, Allgemeines*, ed. Manfred Salzmann (Weimar, Germany: Stadtmuseum Weimar, 1980), 38.

4. Ibid.

5. Schwarze, *The GDR Today*, 57.

6. Franz von Nesselrode, *Germany's Other Half: A Journalist's Appraisal of East Germany* (London: Abelard-Schuman, 1963), 110.

7. Rudolph, "Wirtschaft," 38.

8. Nesselrode, *Germany's Other Half*, 112.

9. "Dokumente der SED, Bd. IV," Berlin 1954.

10. Nesselrode, *Germany's Other Half*, 112.

11. Ibid., 113.

12. *Göttern Chronik*, 1958.

13. In regard to farmer-worker relations and SED expectations about them, see Klemm et al., *Agrargeschichte*, 148–150, 152, 172.

14. James Bennet, "G.M. Success in an Unlikely Place," *New York Times*, October 31, 1994.

CHAPTER 12

1. Jurgen K. A. Thomaneck and James Mellis, *Politics, Society, and Government in the German Democratic Republic: Basic Documents* (Oxford: Berg, 1989), 16.

2. Inheritance contract, December 15, 1967, in possession of the Schorcht family.

3. Hanns Werner Schwarze, *The GDR Today: Life in the "Other" Germany*, trans. John M. Mitchell (London: Wolff, 1973), 56. See also Jörg Roesler, "Die Produktionsbrigaden in der Industrie der DDR. Zentrum der Arbeitswelt?" in Kaelble, Kocka, and Zwahr, *Sozialgeschichte der DDR*, 144–164.

4. David Childs, *The GDR: Moscow's German Ally* (London: Allen and Unwin, 1983), 155.

5. For further information on the comparative value of the mark in the GDR, see Stephen R. Burant, ed., *East Germany: A Country Study* (Washington, DC: Government Printing Office, 1988), 154–155.

6. Mike Dennis, *German Democratic Republic: Politics, Economics, and Society* (London: Pinter, 1988), 140–141.

7. Rudolph, "Landwirtschaft," in *Der Landkreis Weimar. Eine Heimatkunde. Erster Teil: Allgemeines*, ed. Manfred Salzmann (Weimar, Germany: Stadtmuseum Weimar, 1980).

8. *Zehn Jahe Deutschlandpolitik. Die Entwicklung der Beziehungen zwischen der Bundesrepublik Deutschland und der Deutschen Demokratischen Republik 1969–1979: Bericht und Dokumentation* (Bonn: Das Bundesministerium, 1980), 43–44; for the pertinent treaties, see 199–200.

9. See Ronald D. Laing and Aaron Esterson, *Sanity, Madness and the Family* (Harmondsworth, UK: Penguin, 1970), for a general discussion of the pathological consequences of the disconfirmation of daughters by family members.

CHAPTER 13

1. Herr Hertel, personal communication, Kommunalaufsicht des Kreises Weimar.

2. Hermann Weber, *Geschichte der DDR* (Munich: Deutscher Taschenbuch Verlag, 1985), 205–206.

3. David Childs, *The GDR: Moscow's German Ally* (London: Allen and Unwin, 1983), 119.

4. Herr Hertel, personal communication.

5. *Genosse* (comrade) has been a term of address since 1879 among Marxists. Edgar refers to SED members as *Genossen*. *Kollege* (colleagues) means "co-worker." Edgar was addressed as "colleague" by others.

6. See Hermann Weber, *Geschichte der DDR* (Munich: Deutscher Taschenbuch Verlag, 1985), 182, and Henry Krisch, *The German Democratic Republic: The Search for Identity* (Boulder, CO: Westview Press, 1985), 40. See also Jurgen K. A. Thomaneck and James Mellis, *Politics, Society, and Government in the German Democratic Republic: Basic Documents* (Oxford: Berg, 1989), 33–36, 70–72.

7. Thomaneck and Mellis, *Politics, Society, and Government*, 48–49, 75–76.

8. Stephen R. Burant, ed., *East Germany: A Country Study* (Washington, DC: Government Printing Office, 1987), 123–126.

9. For discussion of how shortages came into being, see Katherine Verdery, *What Was Socialism, and What Comes Next?* (Princeton, NJ: Princeton University Press, 1996), 221–222.

10. Ibid., 22.

11. See Article 51 of the *Verfassung der Deutschen Demokratischen Republik*, October 7, 1949; see also *Gesetzblatt der Deutschen Demokratischen Republik*, no. 88 (August 11, 1959).

12. Herr Hertel, personal communication.

13. Thomaneck and Mellis, *Politics, Society, and Government*, 71.

14. Burant, *East Germany*, 125–126.

15. Weimar County Plan, 1969, 17.

16. Lutz Niethammer, "Prolegomena zu einer Geschichte der Gesellschaft," in Kaelble, Kocka, and Zwahr, *Sozialgeschichte der DDR*, 102–105.

17. For information on the Stasi, see Clemens Vollnhals, "Das Ministerium für Staatssicherheit: ein Instrument totalitärer Herrschaftsausübung," in Kaelble, Kocka, and Zwahr, *Sozialgeschichte der DDR*, 498–518; see also Mary Fulbrook, *Anatomy of a Dictatorship: Inside the GDR, 1949–1989* (New York: Oxford University Press, 1995), 46–52.

18. *Chronik Göttern*, clippings, 1972–1984.

19. Walter Venus, "Message in the Belltower," *Chronik Göttern*, December 6, 1931.

CHAPTER 14

1. While attempts to ascribe the diffusion of the nuclear family to the influence of industrialization have been largely refuted by Laslett, Lasch, Hareven, and others, some scholars still argue for confining extended families to areas peripheral to the modern world. See Arthur S. Alderson and Stephen K. Sanderson, "Historic European Household Structures and the Capitalist World-Economy," *Journal of Family History* 16, no. 4 (1991): 419–432.

CHAPTER 15

1. "Magdala," in *Der Landkreis Weimar. Eine Heimatkunde. Erster Teil: Allgemeines*, ed. Manfred Salzmann (Weimar, Germany: Stadtmuseum Weimar, 1980), 20.

2. "Rede von Margot Honecker auf dem Festakt des Ministerrates," *Neues Deutschland*, June 13, 1980; Henry Krisch, *The German Democratic Republic: The Search for Identity* (Boulder, CO: Westview Press, 1985), 159. See Freya Klier, *Lüg Vaterland: Erziehung in der DDR* (Munich: Kindler, 1990), 102–110, for further information about the development of the polytechnic school.

3. Ernst Thälmann Pioneers were named in honor of Ernst Thälmann, leader of the German Communist Party during the final years of the Weimar Republic. From 1933 to 1944, Thälmann was imprisoned by the Nazis. He was murdered in Buchenwald in August 1944. For further information about the pioneers, see *U.S. Office of Education Bulletin*, no. 26 (1959): 112–116.

4. For further information on the Free German Youth (Freie Deutsche Jugend), see Mary Fulbrook, *Anatomy of a Dictatorship: Inside the GDR, 1949–1989* (New York: Oxford University Press, 1995), 60, 94, 166; see also *U.S. Office of Education Bulletin*, 109–111.

5. Henry Krisch, *The German Democratic Republic: The Search for Identity* (Boulder, CO: Westview Press, 1985), 154.

6. For additional information on the *Jugendweihe*, see Fulbrook, *Anatomy of a Dictatorship*, 95–97, 101–102. See also Krisch, *The German Democratic Republic*, 123, 155.

7. Grading in the polytechnical school ran from 1 to 5: 1, very good; 2, good; 3, satisfactory; 4, unsatisfactory; 5, failure. Grades took not only intellectual but also ideological preparation into account. As grades were the sole basis for admission to the university, only a very few "1s" were permitted per class. At the university level, evaluation ranged from 1 to 5 as well. Students who received grades below satisfactory were, on occasion, required to join a study collective (Ute Brandes, personal communication).

8. For information on the twelve-year school, EOS, see Krisch, *The German Democratic Republic*, 161, and David Childs, *The GDR: Moscow's German Ally* (London: Allen and Unwin, 1983), 172–173.

9. In 1968, the Prague Spring attempted reform against the Communist rule, prompting Russia to invade to ensure the stability of the Soviet Bloc.

CHAPTER 16

1. Title of Heidruns's *Diplomarbeit*: "Zur Entwicklung der Hochschulen der USA siet dem Zweiten Weltkrieg unter besonderer Berücksichtigung der Studienorganization und der Studienanforderungen."

2. Most especially, *Economic and Philosophic Manuscripts of 1844* (Berlin: Mega, 1932).

3. Mary Fulbrook, *The Fontana History of Germany, 1918–1990: The Divided Nation* (Fontana: London 1991), 231.

CHAPTER 17

1. Henry David Thoreau, *Civil Disobedience* (Bedford, MA: Applewood Books, 2000).

2. Eugene Genovese, professor of history at Rochester University, is the author of *The Political Economy of Slavery: Studies in the Economy and Society of the Slave South* (New York: Pantheon, 1965), and *Roll, Jordan, Roll: The World the Slaves Made* (New York: Pantheon, 1974), among other titles.

3. Heidrun's dissertation: "*Politisch-Ideologische und Strategisch-Taktische Grundpositionen der Demokratischen Studentbewegung in den USA in den Sechziger Jaren.*" Dissertation zur Erlangung des akademischen Grades eines Doktors des Wissenschaftszweiges der Gesellsschaftswissenshaftlichen Fakultät des Wissenschaftlichen Rates der Friedrich-Schiller-Universität Jena.

4. The Berrigan brothers were Catholic priests and prominent peace activists who were known for protesting the Vietnam War; one was charged with trespassing and tampering with a missile silo.

CHAPTER 18

1. For detailed information on GDR censorship practice in the Anna Amalia Library, Weimar, see the exhibit catalog, *Der rote Punkt*, published by Stiftung Weimarer Klassik, pertaining to the censorship exhibit mounted in Weimar and elsewhere between 1992 and 1994. For more general information concerning censorship in the library system of the GDR, see Jean Edward Smith, *Germany beyond the Wall: People, Politics . . . and Prosperity* (Boston: Little, Brown, 1969), chap. 8.

2. For GDR constitutional guarantees for women, see "Basic Rights and Duties" in Jurgen K. A. Thomaneck and James Mellis, *Politics, Society and Government in the German Democratic Republic: Basic Documents* (Oxford: Berg, 1989), 110–111.

3. Michael Geisler was a member of the Foreign Languages and Literatures Department at MIT from 1979 to 1989. He is currently chair of the Department of German at Middlebury College.

4. Mikhail Gorbachev, *Perestroika: New Thinking for Our Country and the World* (New York: Harper and Row, 1987), 60–63.

5. For further information on Neues Forum, see Dieter Grosser, "The Dynamics of German Unification," in *German Unification: The Unexpected Challenge* (Oxford: Berg, 1992), 14–15; see also Mary Fulbrook, *Anatomy of a Dictatorship: Inside the GDR, 1949–1989* (New York: Oxford University Press, 1995), 242–247.

6. Diana Loesser and Heidrun Schorcht, *Peace Movements in Britain and the USA* (Jena: 1989); Heidrun Schorcht, "Right Is of No Sex"; Frederick Douglass, "Engagement für die Rechte der Frau," in *Wissenschaftliche Zeitschrift der Friedrich-Schiller-Universität Jena*, 489–492 (Jena: 1990); Heidrun Schorcht, "Was ist 'Nationale Sicherheit'?—Bemühungen der USA—Friedensbewegungen einer Neubestimmung der Begriffe," in *Wissenschaftliche Beiträge der Friedrich-Schiller-Universität* (Jena: 1990), 35–53.

7. Ursula Schröter, "Ostdeutsche Frauen im Transformationsprozeß: eine soziologische Analyse zur sozialen Situation ostdeutscher Frauen (1990–1994)," in *Aus Politik und Zeitgeschichte: Beilage zur Wochenzeitung Das Parlament*, B 20, 1995, 31.

8. For example, the paper by Ursula Schröter noted above; also Dorothy J. Rosenberg, "Learning to Say 'I' Instead of 'We': Recent Works on Women in the Former GDR," in *Women in German Yearbook 7* (Lincoln: University of Nebraska Press, 1991).

9. Schröter, "Ostdeutsche Frauen im Transformationsprozess," 41.

10. Hermine G. De Soto, "Equality/Inequality: Contesting Female Personhood in the Process of Making Civil Society in Eastern Germany," in *The Curtain Rises: Rethinking Culture, Ideology, and the State in Eastern Europe*, ed. Hermine G. De Soto and David G. Anderson (Atlantic Highlands, NJ: Humanities Press, 1993), 296.

11. Stephen R. Burant, ed., *East Germany: A Country Study* (Washington, DC: Government Printing Office, 1987), 102.

12. Hildegard Maria Nickel, "Women in the GDR: Will Renewal Pass Them By?" in *Women in German Yearbook 6* (Lincoln: University of Nebraska Press, 1991), 99.

13. Ibid., 100.

14. Ibid., 102–103.

15. De Soto, "Equality/Inequality," 297.

16. Nickel, "Women in the GDR," 99.

17. Mary Fulbrook, *The Fontana History of Germany, 1918–1990: The Divided Nation* (Fontana: London 1991), 239–240.

18. Ibid., 240–241.

19. Schröter, "Ostdeutsche Frauen im Transformationsprozeß," 41.

20. Heidrun's valorization of her relationship to the GDR would seem to be at variance with those who argue for "feminist politics to confront the disadvantages of GDR socialism for women," disadvantages that made difficult the development of critical awareness. ("Feminist *Germanistik* after Unification: A Postscript from the Editors," in *Women in German Yearbook 6*, 109). What is at issue here is not whether East German women were ill-prepared to act on their own behalf in the context of a market economy—most of them were undoubtedly so, Heidrun included—but whether they came to grips with the hegemonic control that SED authoritarianism had imposed on them as women. By 1990, Heidrun had become disaffected with the SED and its obsession with power, yet not necessarily from a feminist perspective. The quarrel that Heidrun

had with the party derived from her sense of its betrayal of socialist ideals. When she resigned from the SED in the spring of that year, Heidrun believed she had expunged her political accounts with the GDR, but not her indebtedness to the social advantages that it had bestowed on her as a person.

CHAPTER 19

1. This obelisk stands opposite the Minute Man statue on the eastern end of the Old North Bridge; it was erected in 1836, at a time when there was no bridge at the site, by the residents of Concord. On Independence Day, 1837, the memorial was dedicated at an event for which Ralph Waldo Emerson wrote his "Concord Hymn." The first, and best known, of the four stanzas of this poem is "By the rude bridge that arched the flood, / Their flag to April's breeze unfurled, / Here once the embattled farmers stood / And fired the shot heard round the world."

CONCLUSION

1. Inheritance contract, August 18, 1992, in possession of the Schorcht family.

2. It was a question not just of West Germany's adjusting to the demands of capitalism but also of restraining them in favor of creating a social democratic welfare state, a goal achievable during the strong economic growth of the seventies. Now that retrenchment is in order, the East is not able to benefit, as it would have earlier, from a more "humane" capitalism.

3. Yolanda Moses and Sue Estroff, "The Futures of Anthropology: Practice, Imagination, and Theory in the 21st Century," *Anthropology Newsletter* 38 (January 1997): 7.

4. George Kateb, *The Inner Ocean, Individualism and Democratic Culture* (Ithaca, NY: Cornell University Press, 1992), 222.

5. Ibid., 225–226.

6. Cornel West, "George Kateb—the Last Emersonian?" in *Liberal Modernism and Democratic Individuality: George Kateb and the Practices of Politics*, ed. Austin Sarat and Dana R. Villa (Princeton, NJ: Princeton University Press, 1996), 43.

7. This is not to say that personal experience cannot be ideologically informed. A case in point is the idealized Nazi belief in *Volksgemeinschaft* (national community), which may have influenced Edgar's own vision of a perfectible Göttern community in the GDR years.

8. Omer Bartov, *Murder in Our Midst: The Holocaust, Industrial Killing, and Representation* (New York: Oxford University Press, 1996); Bartov, *The Eastern Front, 1941–45: German Troops and the Barbarisation of Warfare* (New York: St. Martin's Press, 1986); Bartov, *Hitler's Army: Soldiers, Nazis, and War in the Third Reich* (New York: Oxford University Press, 1991); Christopher R. Browning, *Ordinary Men: Reserve Police Battalion 101 and the Final Solution in Poland* (New York: HarperCollins, 1992); and Hannes Heer and Klaus Naumann, eds., *Vernichtungskrieg: Verbrechen der Wehrmacht 1941 bis 1944* (Hamburg: Hamburger Edition, 1995).

9. Margarete Mitscherlich, "Unified Germany: Stabilizing Influence or Threat?" *Partisan Review* 62, no. 4 (1995): 527–534.

Index